Degrees That Matter

Urban and Industrial Environments
Series editor: Robert Gottlieb, Henry R. Luce Professor of Urban and Environmental Policy, Occidental College

For a complete list of books published in this series, please see the back of the book.

Degrees That Matter

Climate Change and the University

Ann Rappaport and Sarah Hammond Creighton

The MIT Press
Cambridge, Massachusetts
London, England

MIT Press books may be purchased at special quantity discounts for business or sales promotional use. For information, please e-mail special_sales@mitpress.mit.edu or write to Special Sales Department, The MIT Press, 55 Hayward Street, Cambridge, MA 02142.

This book was set in Sabon by SNP Best-set Typesetter Ltd., Hong Kong. Printed and bound in the United States of America.

Printed on recycled paper.

Library of Congress Cataloging-in-Publication Data

Rappaport, Ann.
Degrees that matter : climate change and the university / Ann Rappaport and Sarah Hammond Creighton.
p. cm.—(Urban and industrial environments)
Includes bibliographical references and index.
ISBN: 978-0-262-18258-4 (hardcover : alk. paper)—978-0-262-68166-7 (pbk. : alk. paper)
1. Climatic changes—Environmental aspects—United States. 2. Greenhouse gas mitigation—United States. 3. Global environmental change—United States. 4. Science—Study and teaching—United States. 5. Universities and colleges—United States. I. Creighton, Sarah Hammond. II. Title.
QC981.8.C5R367 2007
363.738′746071173—dc22

2006030105

10 9 8 7 6 5 4 3 2

To William R. Moomaw, friend and mentor.
Bill's generous spirit and creativity have inspired generations of students. His willingness to take action to protect the planet is an inspiration to us all.

Tables, Figures, and Boxes

Action on climate change can and should be taken with all of these communities working collaboratively to leverage their resources. Individuals, institutions, businesses, nongovernmental organizations, foundations, governments, and intergovernmental organizations all have something to contribute to the effort. Colleges and universities bear a special responsibility to facilitate interaction among these communities by providing scientific knowledge, technological innovation, and the next generation of leaders.

Colleges and universities also have an opportunity to lead by example, by taking action to reduce our own contributions to global warming. As steward of a complex physical community, a university president can exercise leadership in reducing the climate implications of a wide range of decisions related to resource use. Implementing these decisions requires the innovation, practical knowledge, and hard work of our institutions' best minds, be they faculty, staff, or students. Actions include using energy that emits no heat-trapping gases, and constructing and renovating buildings so that they are extremely energy efficient. Physical planning for campuses should have goals related to sustainability and must explore the implications of a changed climate so that campus infrastructure can be protected. Plans must also place a high priority on dramatic reduction of greenhouse gas—not just slowed emissions growth. As a leader of a physical community, a university president can also establish norms for personal actions that have climate change implications. We can use the university as a learning laboratory, engaging students, staff, and faculty to take climate action.

This book focuses on colleges and universities because actions on our campuses have unique value in society: they have a built-in multiplier effect. At Tufts University, we are educating our students to become active, engaged, *effective* citizens across diverse disciplines. Our goal is to nurture a community of

- People comfortable dealing with ambiguity
- People willing to take a risk to make a difference
- People more interested in solving problems than in taking credit
- People who are both effective advocates and aggressive listeners
- People who are eager to imagine and implement large, daring, multifaceted solutions—together

We believe that these attributes are essential to improving society and that even a problem as profound as climate change can be addressed by people who are engaged and active. Colleges and universities can help lead the way.

Lawrence S. Bacow
President
Tufts University

Acknowledgments

We gratefully acknowledge the efforts of many people who made this book possible. Friends and colleagues at Tufts University are responsible for many of the ideas and climate actions we describe. We are fortunate to have the active participation and support of Tufts' president, Lawrence Bacow. Bill Moomaw conceived Tufts Climate Initiative in the 1998–1999 academic year with the endorsement of John DiBiaggio, who was Tufts' president at the time. Without the collaboration of many people in the Division of Operations, progress in slowing global warming on our campus would be impossible; they include Betsy Isenstein, John Roberto, John Vik, Dawn Quirk, Elliott Miller, Sheila Chisholm, Bob Bertram, and Dick Goulet.

Faculty whose student projects inform this work include Maria Flytzani-Stephanopoulos, Kent Portney, Bill Moomaw, Chris Swan, and Jeff Zabel. Countless students have educated us through projects and coursework. Those whose work or stories are reflected in these pages include Sinan Seyhun, Nicole Robillard Wobus, Jennifer Baldwin, Robin Struwe, Kristen Marcell, John Larsen, Jon Grosshans, and Nate Phelps. Particular thanks to the students in Ann Rappaport's course on climate change policy, planning, and action.

Those who have provided funds for the Tufts Climate Initiative deserve special mention. They include the Henry P. Kendall Foundation, the Rockefeller Brothers Fund, the Jonathan M. Tisch College for Citizenship and Public Service, Tufts Institute for the Environment, and Toyota Motor Company USA.

This book reflects insightful comments from many thoughtful people, including Bill Moomaw, Anja Kollmuss, Ramsay Huntley, Janice Snow, and anonymous reviewers. Special thanks to Kollmuss, whose steadfast offer curricular reflections based on approaches we and our colleagues have taken.

Because students remain on campus for a relatively short time, their efforts devoted to climate action can be lost to the community when they graduate. Students generate a level of enthusiasm, passion, and excitement that gets many new projects off the ground. We offer a detailed conceptual map to tie the pieces together across groups and across time—for example, as students transition to alumni and colleges make connections to keep them involved.

Given the considerable variation across academic institutions, it can be a challenge to determine who needs to be involved in reducing the organization's contribution to climate change. As a result, we focus on understanding an institution's decision-making framework. Whatever the model, people are at the center of successful climate change action measures; the more people understand climate action, the more rapidly progress can be made. We provide an introduction to the technical knowledge you need in order to communicate effectively with professionals in the trades to develop the most efficient climate change mitigation solutions tailored to your campus. We also provide additional references and resources to expand your climate action knowledge.

Why We Must Focus on Climate Change and Do It Now

Our Arctic neighbors have sent us clear images of what is in store for the rest of the planet if we delay. Climate change demands attention. The unwillingness of political leaders in either major political party in Washington to ratify the Kyoto Protocol was the catalyst for the initial commitment at Tufts, but it is not the sole factor sustaining the effort. The energy crisis of the 1970s motivated many colleges and universities to take aggressive efficiency measures, many of which were subsequently relegated to the dustbin of institutional memory as fuel costs decreased. Energy costs and reliability are again a concern for institutional decision makers; technological developments since the 1970s have brought many new long-term operational savings. For the university of the future, lower energy costs and increased reliability can be a significant competitive advantage.

Climate change is inherently different from regional and local environmental problems such as water quality and waste disposal. Water and waste solutions can be developed in communities, regions, or watersheds where the people who invested in resource protection can observe and measure the results. But climate-altering gases are international travelers. Emissions contribute equally to global climate change regardless of where they are released, just as ozone-depleting chemicals did by opening holes in the ozone layer of the stratosphere. The ozone-depletion problem involved a relatively small number of human-made chemicals produced by a limited number of manufacturers. When DuPont, the largest manufacturer, announced ambitious plans to scale back and then eliminate the production of ozone-depleting chemicals on the target list, international consensus on phaseout timetables soon emerged and was reflected in the Montreal Protocol. In contrast to ozone, the most prevalent gases that contribute to climate change are associated with ubiquitous processes such as combustion (carbon dioxide), decay (methane), and agriculture (methane and nitrous oxide). This means that we all generate greenhouse gas, whether we are subsistence farmers burning wood for cooking fuel, office workers in an insurance company, parents driving children to soccer games, or executives at General Motors.

Decision makers in many companies take climate change seriously. In a 1998 survey of Fortune 500 companies, Ann Rappaport and John Blydenburgh[8] asked environment personnel what they believe will be the greatest environmental challenge for their company over the next ten years. Climate change was the most frequently selected response (25.5 percent), followed by sustainable development (15.3 percent). Responding to both climate change and sustainable development requires companies to engage in long-term planning. Often companies are criticized for the short-term focus of their financial and environmental strategies. More recently, companies such as Swiss Re have been articulating the link between climate change and the long-term financial viability of a wide range of industries.[9] Climate change also has clear implications for the investment portfolios of universities, colleges, and other institutions.[10]

Research reveals that Americans believe climate change exists, but that many people do not understand that climate change is human-made and that there are solutions available to reduce their contribution. Few people understand that burning fossil fuels is the most important contributor to climate change,[11] or make the connection that most of the electricity in the United States is generated by burning fossil fuels. Student surveys at Tufts reveal some of the same misunderstandings.

Academic Research and Teaching for Climate Action

Clearly an educational effort is needed to frame effective efforts that yield constructive action; academia is an ideal place to develop and test strategies. However, educational efforts must move well beyond the classroom, and quickly, to reach the full range of climate change decision makers.

Although the complexity of climate change and the multiplicity of climate-altering gas sources presents an enormous challenge, such complexity is tailor-made for academic inquiry and knowledge-inspired action across disciplines from engineering (designing highly efficient motors, developing renewable power technologies or improved energy systems) to the humanities (examining the nature of our generation's obligations to manage resources in a way that does not compromise future generations). Crafting effective and efficient actions is also a much more complex challenge than classic antipollution campaigns and slogans, such as "reduce, reuse, recycle." Sophisticated systems approaches are needed to identify actions that minimize climate-altering gas emissions and maximize returns on investments.

A great deal of academic and media criticism has been focused on the Kyoto Protocol, providing rich source material for courses on climate change and environmental stewardship. There is a vast difference between debating approaches, as politicians are still doing, and taking action, which is what we all can and should be doing.

Where to Begin: Establishing a Baseline

The Kyoto Protocol uses 1990 emissions as a baseline, so the first step is to quantify 1990 emissions, current emissions, and growth projections, so that emission reduction targets can be established. This is conceptually clear and simple; however, moving from idea to action is not so easy. Participating in the development of an emissions inventory for an organization such as a university illustrates how challenging it can be to execute such a simple concept in real life. And if it is difficult to develop an inventory for a reasonably well-managed university community, the implications for conducting an emissions inventory in a large, diverse industrial or developing country immediately become clear. Academic institutions are well placed to design and implement better tools and strategies.

Once an emissions inventory has been developed, it is then possible to evaluate progress toward quantitative goals on a regular basis. The fact that greenhouse gas reductions lend themselves so well to evaluation is another substantial asset of embracing climate change. Emissions can be tracked on several levels, informing both campuswide and individual project decisions. Actions taken to reduce greenhouse gas emissions are more costly than measures typically taken as part of campus greening; as a consequence, it may be important to establish a sound fiscal basis for investments. With relatively little effort it is possible to calculate payback periods for activities such as lighting efficiency improvements, and to compare the costs per unit of carbon reduced for most alternative program approaches. The ability to evaluate progress is a significant educational and managerial asset.

Linking Research and Action

Climate change research and action span social, environmental, and economic systems. Research to refine our understanding of global warming's consequences and action to address these problems can and must be carried out simultaneously. Climate change has been the subject of thoughtful and rigorous scientific inquiry, extensive diplomatic negotiations, and shameless political manipulation. Political leaders in Washington have delivered unclear and confusing messages as to what can be done about climate change.

There is a false notion in the public discourse that taking unilateral action on climate change is a foolish waste of money. In fact, many energy-efficiency measures applied to buildings will yield operational cost savings that will accrue over the entire operational life of a building, which for most colleges and universities will be many decades to a century or more. The public discourse has been incomplete because it fails to pose and examine very important questions:

- What will be the impact on us of reducing emissions?
- How will our lives be different if we take measures now that benefit future generations?
- Might these measures also confer benefits on us in the short term?

support university purchase of wind power. An error in producing the ballot left the measure off the regular student government ballot. A delay until the next regular election in the fall had some strategic merit, but one of the prime movers was a senior who would be graduating.

A makeup vote just for the wind petition was scheduled, but wind-power advocates were given only two days' notice. And the vote would be during reading period before final exams. Who would pay attention? How could advocates possibly get the 25 percent participation required for a valid vote? On voting day as advocates planted hundreds of orange, green, and yellow pinwheels on the academic quad a cold drizzle was falling. Would anyone care? When voting closed at midnight, 40 percent of the students had participated with 88 percent voting for the wind initiative. Equally important was the 40 percent voter turnout, higher than participation in the regularly scheduled student government election. Is it possible that interest in climate action will engage students more broadly in political participation and local and national governance?

In the final chapter of this book we reflect on what inspires action and what makes a credible effort by colleges and universities committed to climate action. We argue that only by going well beyond business as usual can institutions legitimately claim that they are acting responsibly and educating future generations for a world transformed by global warming.

Summary

Colleges and universities, like most people, organizations, and companies, contribute to global warming while conducting normal activities. People are increasingly concerned about climate change because they understand that the consequences for human well-being and for ecosystems may be dramatic. Yet few people know how to transform their concern into action. Through education and innovation, colleges and universities have a unique opportunity to lead the transformation. Colleges and universities can reduce their emissions of heat-trapping gas at the same time they educate and inspire members of their communities, develop new technologies and, in many cases, reduce long-term energy costs.

2

Climate Change Basics and the University

How do colleges and universities contribute to climate change? This chapter presents information on climate change linked to action at colleges and universities. Advocates for climate action will frequently find themselves providing explanations of global warming to people from varying backgrounds with a wide range of exposure to the issue. We present a very brief overview of the science and an equally brief glimpse at just a few of the political issues that surround climate change action. Heat-trapping gas emission sources, particularly at colleges and universities, are also explored briefly; details are included in subsequent chapters.

We use emission data from colleges around the country to examine some of the implications of comparing our emission profiles with those of others. One of the issues we explore is the relationship between wealth and emissions. In the next chapter, we talk about generating a campus emission profile with an inventory.

Climate Change Basics for Decision Makers and Advocates

Climate change or global warming refers to the changes in the earth's temperature that are associated with human activity. We generally use the term *climate change* because it incorporates the range of outcomes beyond rising temperatures. Even though the global average temperature will rise, some areas will become cooler. We know that some people prefer the term *global warming* because it sounds more urgent. Both terms refer to the same phenomenon. The basic mechanism is frequently captured by the idea of a "greenhouse effect"; however, a more familiar analogy is that of a "hot car."

The atmosphere allows visible solar radiation to reach the earth relatively easily. The atmosphere absorbs the infrared radiation emitted by the earth's surface and radiates it back to the earth in much the same way a greenhouse or the windows of a car trap heat as the sun's rays pass through the glass, and the radiant heat generated does not pass back through the glass. The "greenhouse effect" causes the surface of the earth to be much warmer that it would be without the atmosphere. Without some greenhouse effect, life as we know it would not exist on earth.[1] The problem is that since the industrial revolution, there has been an enormous increase in the atmosphere of gases that trap heat, and current levels are greater than at any point in the last 650,000 years.[2]

Climate-altering gases include water vapor, carbon dioxide, and methane, all of which trap heat and affect the earth's surface temperature.[3] These gases come from a wide variety of sources, both natural and anthropogenic. Water vapor is the primary heat-trapping gas in the atmosphere, and carbon dioxide (CO_2) is the major heat-trapping gas from human activities. Carbon dioxide results primarily from the combustion of fossil fuels for energy (electricity generation, heating, cooling) and transportation.

Definitions of climate change usually make a clear distinction between changes in the earth's climate that are associated with human activities and changes that are natural. For example, climate change, defined by the Framework Convention on Climate Change, "refers to a change of climate which is attributed directly or indirectly to human activity that alters the composition of the global atmosphere and which is *in addition to natural climate variability* observed over comparable time periods."[4]

Projections for temperature increases over the future lifetimes of current students are an additional 3°F (Fahrenheit) to 7°F depending on societal choices.[5] Lower increases are possible if sufficient changes in emissions are made. To put these numbers in perspective, the most recent ice age was about 6°F cooler than current average temperatures, so small changes in global average temperature can have enormous implications for life as we know it.[6]

Not only do small changes in global average temperature portend large changes for the earth's inhabitants, but the climate system has a long delay. For example, if we were to make an abrupt change, stopping all anthropogenic emissions of CO_2 tomorrow, models show that global average temperature will continue to rise for about thirty years, and then will fall very slowly.[7] Many people think intuitively that if you stop emissions of heat-trapping gas today, temperatures will start to fall tomorrow, but the system delay means that temperatures will not start falling until our children or grandchildren are in decision-making positions. The work of John D. Sterman and Linda Booth Sweeney shows that the implications of delay in the climate system are not understood by highly educated graduate students, so it is little wonder that some politicians believe incorrectly that a "wait and see" approach makes sense.[8] In fact, each day we delay action, we have a larger problem to solve. Sterman and Sweeney note that "when there are long time delays between actions and their effects, as in the climate system, wait and see can be disastrous."[9]

Talking about Climate Change

Although these climate change basics have appeared in the popular press and in government and scholarly publications, a challenge in effectively presenting this information to decision makers persists. The challenge includes the need to simplify complicated concepts and to provide enough information to motivate appropriate action. When talking to general audiences, some people use the analogy of a "heat-trapping blanket" to explain the effect of climate-altering gases on the earth's surface temperature. Scientists feel that this image rests on incorrect science and is therefore misleading. The example of a "hot car" is viewed as more accurate than a heat-trapping blanket, and more familiar and thus more effective than the analogy of a greenhouse. Box 2.1 presents one approach to communicating climate change. A challenge to readers: Can you improve on these messages with an approach that is both accurate and effective with general audiences?

Governments and Climate Change

Scientists are playing an increasingly active and vocal role in talking with government decision makers about the adverse impacts associated with continued high levels of carbon emissions. A great deal of the interaction between scientists and governments is based on work by the Intergovernmental Panel on Climate Change (IPCC). The IPCC is composed of over 2,500 experts, many from academic institutions around the developed countries would have to decrease theirs, dramatically in some cases.[15] Because developed countries dominated the discussion in Kyoto, it led to several features that favor industrialized nations. For example, the Kyoto agreement allows developed countries to take credit for emission reductions in developing countries through the Clean Development Mechanism (CDM). The assumption is that these reductions will be less costly than reductions in developed countries. Developing countries may benefit from technology transfers under CDM. But when developing countries eventually have to reduce their emissions, they will be disadvantaged if all of the inexpensive emission reductions have been taken by developed countries under CDM.

Even the best international agreements have limitations that are not widely understood. International agreements are agreements among governments that each will create a legal infrastructure to ensure that its country's commitments are upheld. For example, the United States implemented the Montreal Protocol by amending the Clean Air Act to restrict the manufacture and use of ozone-depleting chemicals, and each country ratifying the Montreal Protocol took measures with comparable effect. Most international environmental agreements do not have substantive enforcement mechanisms. There is no World Environmental Protection Agency and even if there were, the agency's power would be limited by the overriding concern of countries to preserve their national sovereignty. This means that the strength of an international agreement on climate change really rests on the willingness and ability of the governments and citizens of each participating country to pass effective domestic laws and ensure their enforcement.

The decision of the United States not to ratify the Kyoto Protocol and not to take meaningful regulatory action at the national level to reduce heat-trapping gas emissions has been the source of international consternation. Historically, the United States has taken a leadership role in environmental agreements and it is disappointing to see this nation among the laggards. North America is the highest fossil-fuel, carbon dioxide–emitting region of the world with 1.65 billion tons of carbon in 2000, and the United States accounts for about 93 percent of the North American emissions. This 2000 total is an all-time high for North America and represents a 1.5 percent increase from 1999. Per capita emissions have been consistently high and well above those for any other region.[16] Given the global nature of the problem, the failure of the United States to curb carbon emissions will negate the efforts of others.

Subnational Initiatives for Climate Change

In the absence of national leadership on climate change, several states and local governments within the United States have made ambitious commitments to heat-trapping gas reduction. Climate change advocates at colleges and universities can both support and be supported by regional, state, and local efforts.

Although the United States is a party to the United Nations' Framework Convention on Climate Change, the federal government has only implemented mild voluntary initiatives such as the Environmental Protection Agency's Climate Leaders program.[17] However, this has not prevented many subnational governments (state, regional, and local) and nongovernmental organizations (as well as private institutions) from moving ahead with their own climate change goals. There are numerous examples of action happening throughout the United States. We highlight only a few of the activities here (and it should be noted that this discussion only covers the United States—there are many organizations working around the world on these issues, from the rainforests of the Amazon to the cities of Europe and Asia):

- *ICLEI* The International Council for Local Environmental Initiatives' (ICLEI) Cities for Climate Protection program brings together municipalities across the country in a shared effort to reduce greenhouse gas emissions. Currently, more than 150 cities and counties, from King County, Washington, to New Orleans, Louisiana, to Gloucester, Massachusetts, are members of the campaign. Among the ICLEI Cities for Climate Protection are three of Tufts' host communities: Medford, Somerville, and Boston. Local governments have the opportunity to influence emissions through land-use decisions, local building codes, and government purchasing as well as embodying the dictum to "think globally, act locally."[18]
- *State Attorneys General lawsuits* A diverse group of states have banded together to sue private companies in an effort to demand they reduce their CO_2 emissions. Led by Attorney General Eliot Spitzer of New York, this lawsuit targets companies that collectively produce 10 percent of the nation's emissions. Instead of seeking monetary damages,

the lawsuit asks defendants to decrease their emissions of greenhouse gases. This is the first lawsuit to target specific polluters.[19]

• *New England Governors/Eastern Canadian Premiers Climate Action Plan* In 2001, the New England Governors/Eastern Canadian Premiers (NEG/ECP), a group of six states and five provinces, agreed on a climate change plan that reduces greenhouse gas emissions from this region of North America. This plan roughly parallels the goals of the Kyoto Protocol over its first two decades of implementation, but then goes on to call for much larger reductions of 75 to 85 percent in future decades (as dictated by science, assuming the climate predictions improve over time). This plan is innovative in that in addition to climate change goals, it is tied to other regional goals such as preservation of open space and promotion of the region's high-tech industries.[20]

The Role of Wealth

One of the arguments advanced in the United States against ratifying the Kyoto Protocol is that it allows developing countries continued growth in emissions while requiring developed countries to reduce greenhouse gas emissions. Although it is not our intent to analyze fully this debate, it is useful to consider current greenhouse gas emissions. There is a clear relationship between countries' wealth and their emissions of climate-altering gases.

Emissions by Countries and Wealth

Table 2.1 provides information on CO_2 emissions in select developed and developing countries and wealth as measured by gross domestic product (GDP). Per capita emissions of heat-trapping gas are lowest in developing countries and greatest in developed countries. In general, greater GDP is associated with greater emissions. However, note the variation among developed countries. Japan, for example, produces far fewer emissions per capita than the United States but has higher per capita GDP. History may offer a partial explanation for this phenomenon in that Japan was forced to rebuild its industrial infrastructure after it was devastated in World War II. Reconstruction allowed Japanese industry to profit from technological and managerial innovations that result in greater efficiencies than in many U.S. companies, particularly those with physical plants dating to the first half of the twentieth century. In

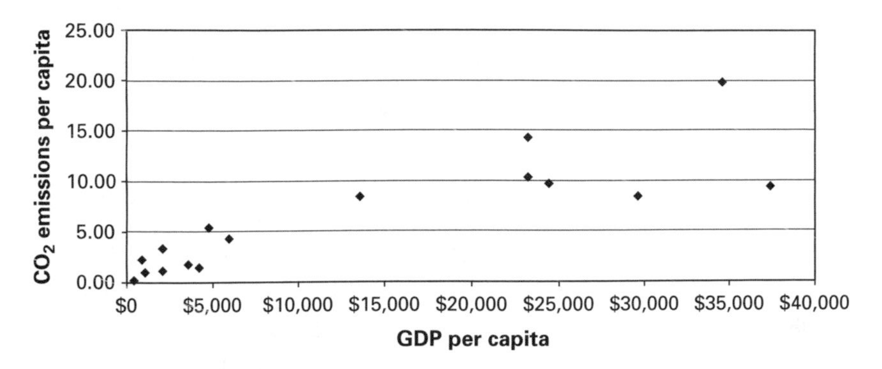

Figure 2.1
Plot of GDP and emission data for select countries

addition, Japan has a high level of public transit use and small homes. But developed countries such as the United Kingdom and Finland also have considerably lower emissions per capita than the United States.

Figure 2.1 shows the data on national wealth and carbon emissions in the form of a plot so that the relationship can be more easily visualized. A detailed analysis of the relationship between national wealth and carbon emissions is outside the scope of this work; however, figure 2.1 does provide the context for exploring the effect of wealth on college and university emissions of greenhouse gases.

Wealth and Heat-Trapping Gas Emissions Generated by Colleges and Universities

Colleges and universities generate heat-trapping gases from several sources, including electricity generation, heating and cooling of buildings, and transport of people and goods. Only a handful of colleges and universities have inventoried their climate-altering gas emissions using a systematic approach. We gathered information from several institutions that have conducted inventories, and even though the number of observations is limited, we see evidence that wealth is a factor in college and university emissions just as it is with countries.

As table 2.2 shows, the institutions with the greatest endowment tend to be those with the greatest impact on climate change. For example, Connecticut College produced about 14,800 metric tonnes carbon dioxide equivalent (MTCDE) in 2000, and Wellesley College produced

Table 2.2
Endowment and emissions information for select colleges and universities

	Emissions (MTCDE)	Emissions /student (MTCE /student)	Endowment × $1,000	Endowment /student
Yale	291,696	6.94	$12,740,896	$1,110,705
Wellesley	43,702	5.16	$1,179,988	$510,375
Middlebury	35,000	3.94	$664,781	$274,250
Cornell	280,000	3.76	$3,200,000	$157,372
Connecticut College	14,834	2.19	$151,927	$82,167
Tufts	88,924	2.55	$771,793	$81,164
Carnegie Mellon	356,100	9.95	$763,717	$78,282
UC Berkeley	150,000	1.24	$2,184,840	$66,077
Vermont	65,800	1.64	$223,865	$20,413
University of New Hampshire	58,062	1.02	$168,692	$10,823
University of Colorado	126,746	1.07	$225,479	$6,954
College of Charleston	38,712	0.92	$34,696	$3,008
Portland State University	30,808	0.36	$19,771	$857

Sources:

Administrative Council, "Meeting Minutes," *Wellesley College*, April 2005, http://www.wellesley.edu/AdminCouncil/meetings.html.

Carnegie Mellon University, *Campus-wide Energy Indicators*, undated, http://www.cmu.edu/greenpractices/facts_figures/energy_consumption.htm.

G. Carroll, *Total Square Footage at the University of Colorado*, personal communication, undated.

"College Briefs: At a Glance," *U.S. News & World Report*, 2006, http://www.usnews.com/usnews/edu/college/tools/brief/cosearch_advanced_brief.php.

College of Charleston's Campus Sustainability, *College of Charleston Emissions*, undated, http://www.cofc.edu/sustainability/ghgemissions.htm.

Cornell University, *Emissions Information*, 2004, http://www.utilities.cornell.edu.

Cornell University, *Endowment*, 2006, http://www.alumni.cornell.edu/endowment.htm.

Cornell University, *Facts about Cornell*, 2006, http://www.cornell.edu/about/facts/stats.cfm.

Table 2.2
(continued)

T. Cruickshank, "GHG Emissions Inventory," University of Connecticut Office of Environmental Policy, undated, http://www.ecohusky.uconn.edu/climatechange.html.

D. Dagan, "A Summary of Energy Consumption and Greenhouse Gas Emissions at Middlebury College," *ES 500 Independent Study, Middlebury College*," fall 2002, http://www.middlebury.edu/NR/rdonlyres/26273E73-9E1D-4283-864C-1C6E9E5B74F9/0/Emissions_InvD_dagan.pdf.

J. Dziubeck, *Connecticut College: Greenhouse Gas Emissions Inventory 1990–2002*, Connecticut College, October 2003, http://camel2.conncoll.edu/ccrec/greennet/GHG_Report.pdf.

Energy Task Force, Yale University, "Engaging the community in Yale's Emission Reduction and Energy Conservation Strategy," undated, http://www.yale.edu/sustainability/Energyforum.pdf.

Institute of Education Sciences, "Enrollment," *Integrated Postsecondary Education Data System*, undated, http://nces.ed.gov/ipeds/data.asp.

Middlebury College, "Standard Eight: Physical Resources," *The President and Fellows of Middlebury College*, undated, http://www.middlebury.edu/administration/secretary/news/reaccreditation/Std_08_Physical_Res_0823.htm.

"NACUBO Endowment Study," *Chronicle of Higher Education*, June 30, 2003, http://thecenter.ufl.edu/research2002.html.

J. Norwell, *Total Square Footage at Connecticut College*, personal communication, 2005.

Office of Institutional Research, College of Charleston, *Planning and Reference Guide 2005*, http://irp.cofc.edu/prg/IRPlanandRefGuide2005.pdf.

Portland State University, *About PSU*, undated, http://www.sustain.pdx.edu/hm_about_psu.php.

Portland State University, *PSU's Energy Usage*, undated, http://www.sustain.pdx.edu/ci_energy_psu_usage.php.

Student Environmental Action Coalition, *Greenhouse Gas Audit for the University of North Carolina at Asheville*, undated, http://www.seac.org/energy/resources/ghgauditunca.pdf.

University of California at Berkeley, *Campus Facilities Service*, undated, http://physicalplant.berkeley.edu/cfs.asp.

University of California at Berkeley, "Energy," *Chancellor's Advisory Committee on Sustainability*, undated, http://sustainability.berkeley.edu/assessment/pdf/CACS_UCB_Assessment_2_Energy.pdf.

University of Colorado Environmental Center, "Carbon Emissions Summary," *University of Colorado at Boulder*, 2005, http://ecenter.colorado.edu/energy/projects/emissions/numbers.html.

University of Massachusetts at Amherst, *All Buildings Sort by SqFt*, 2003, http://www.facil.umass.edu/~utildept/html/projects/EnergyProject/BuildingMeterList.htm.

University of New Hampshire, *Campus Master Plan 2004*, undated, http://www.unh.edu/cmp/pdf/Appendix1.pdf.

Table 2.2
(continued)

University of New Hampshire, *University of New Hampshire, Durham Campus, Greenhouse Gas Emissions Inventory 1990–2000: Executive Summary*, 2001, http://www.sustainableunh.unh.edu/climate_ed/greenhouse-gas-invnt/unh-exec-summary.pdf.
University Planning, Carnegie Mellon University, *Carnegie Mellon Factbook 2005*, vol. 19, February 2005, http://www.cmu.edu/ira/facts2005/2005%20Fact%20Book%20Revised%20Version.pdf.
University of Vermont, "Findings," *Climate Change*, http://www.uvm.edu/climatechange/?Page=Findings.htm.
Wellesley College, *Endowment*, 2005, http://www.wellesley.edu/Resources/why/wellesley.html.
Wellesley College, *It Ain't Easy Being Green: An Audit of Wellesley College's Greenhouse Gas Emissions*, undated, http://cs.wellesley.edu/~weed/papers/ES300-info-sheet.html.
Yale University Office of Institutional Research, *Yale University: History of Buildings Constructed or Acquired, 1717–1999*, 2000, http://www.yale.edu/oir/book_numbers_updated/M8_Building_History_of_Campus.pdf.

about 43,700 MTCDE in the same year. Both institutions experience roughly the same weather (a key factor in determining energy use). Wellesley has more students, but on a per student basis (one logical way to normalize this indicator) the discrepancy is still great: Connecticut College emitted 2.19 MTCE per student in 2000, and Wellesley College emitted 5.16 MTCE per student in the same year. Endowments also reveal a discrepancy: Connecticut College has about $152,000 per student, whereas Wellesley has a per student endowment that is more than seven times larger.

Figure 2.2 shows per capita endowment and emissions data for colleges and universities as a plot. If you compare these data with the per capita emissions and GDP for countries in figure 2.1, the relationship between wealth and emissions appears similar for colleges and universities.

Endowment is not an ideal measure of institutional wealth, but it is an indicator that is used by the educational community to make cross-institutional comparisons (for that matter, GDP is far from a perfect measure of national wealth). The link between wealth and college and university emissions is not a great surprise. Wealthier institutions often have more facilities for research and instruction than their financially

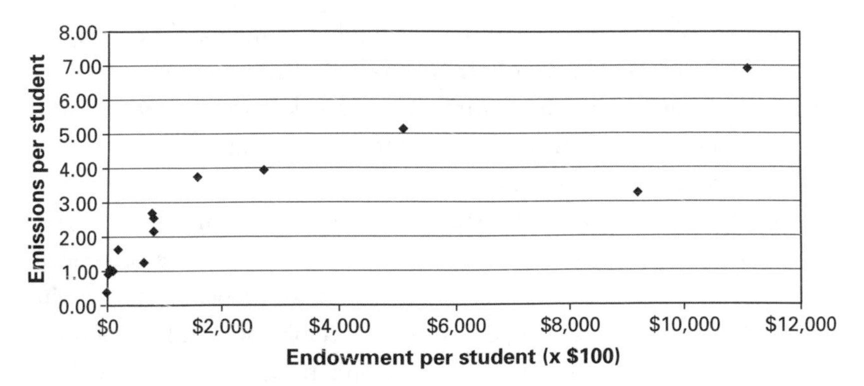

Figure 2.2
Plot of college and university endowment and emission data

strapped counterparts. And more square feet of building generally means more emissions of greenhouse gases.

There are several qualifiers to the information presented in table 2.2 and figure 2.2. For example, not all institutions are providing comparable services; some may provide student housing and some may not, or some may be research-intensive and some may not be, so care must be taken in drawing conclusions. In compiling these data, we used inventory information provided by each institution, typically posted on the college or university website. In cases where there were missing data, such as square feet of buildings, we supplemented with a telephone call to a knowledgeable person on campus. (See the source note for this table for details.) In most cases, we do not know what methods were used by each institution to prepare its inventory. We offer an approach to inventory development in chapter 3 and provide inventory details in appendix C.

The data in figure 2.2, along with our observations, suggest that even if a less affluent institution such as a community college has poorly maintained or inefficient buildings, it is likely to have low greenhouse gas emissions per student. This is because community colleges often provide educational services to a large number of students in a relatively small amount of space and often use their space from early in the morning until late at night. In contrast, a more affluent institution has buildings that may be considerably better maintained and more energy efficient, but the greater amount of space per student and the resulting less intensive space utilization results in higher emissions per capita. Nonetheless, our qualitative observations of a variety of schools show that less affluent schools have often been quicker to embrace routine energy-efficiency measures than those with more resources.

These observations raise interesting questions about how best to compare the climate change impacts of colleges and universities. If the metric selected is energy use per square foot, the list will be ordered in one way, and if the metric is energy use or emissions per student, a different order is certain to emerge. We believe that a per student metric is the most relevant measure of an institution's progress, but normalized metrics have an important limitation. While normalized metrics facilitate comparisons among institutions, it is useful to bear in mind that total emissions are the relevant measure when assessing the impact of the college or university on the environment. In other words, colleges and universities cannot compare their emissions to those of their peers and take comfort in the fact that others are emitting more climate-altering gases. Institutions that wish to be socially responsible will take action to reduce their emissions because they understand that all units of carbon dioxide released to the atmosphere contribute equally to climate change.

Sources of Heat-Trapping Gases

Combustion of fossil fuels is the most common source of carbon dioxide. But not all fuels are equal—for example, bituminous coal emits 205 pounds of carbon dioxide per million Btu, while natural gas emits 117 pounds of carbon dioxide per million Btu. The carbon dioxide implications of fuel choices are shown in appendix D.

Different fuels have different greenhouse gas emission characteristics when burned (for heat or electricity or steam generation), so there are often benefits to changing fuels for select buildings or for the entire campus in order to have lasting benefits for climate change. For example, switching a building's boiler from oil to natural gas will result in fewer emissions of carbon dioxide, since natural gas produces more heat per carbon atom than oil or coal. On campuses where coal is still burned, significant progress toward reducing carbon dioxide emissions can be made by converting these boilers (often central heating and cooling plants) to natural gas. Additionally, increasing the efficiency of the systems themselves at the same time will further increase these benefits. These efficiency benefits can be substantial on campuses using equipment that often is twenty or thirty or more years old. The advantage of these investments in efficiency can be long lasting, since these changes in hardware essentially hardwire the emission reduction in place.

Fuel switching is not limited to institutional heating plants. Fuel switching can also have significant benefit in the generation of electricity. Since most institutions' emissions come, in large part, from electricity generation by utilities, the electric utility's fuel selection will have a dramatic effect on the campus emission profile. When a university can purchase electricity or generate it from fuels that have lower emissions to generate the same power, emission reductions result. In January 2006, Tufts made a switch from the electricity with standard regional fuel mix to a provider whose existing hydropower will eventually generate about 80 percent of its power. The result is that our emissions immediately dropped by about 40 percent. Other institutions are beginning to consider the fuels used to generate their electricity in their purchasing decisions. When wind or hydropower are part of that mix, emissions decrease. Increasing nuclear power in the fuel mix also reduces emissions, but there are significant downsides to that choice.

When an institution purchases electricity or other energy such as steam, the college is usually at the mercy of the provider regarding fuels used and resulting emissions. For example at Tufts, emissions from our Boston campus decreased between 1990 and 1998 because the fuels used for generating electricity became less carbon dioxide intensive, despite the fact that our electricity use during that period increased. However, the company that provides that campus with steam recently changed to more carbon dioxide–intensive fuels, explaining, in part, our emissions increase in 2004. The balance of that increase was because an energy-intensive laboratory building was brought online.

Other Sources of Greenhouse Gas Emissions

While carbon dioxide is the most common source of heat-trapping gas, both worldwide and at colleges and universities, other gases also contribute to climate change. A single molecule of many of these other greenhouse gases has the ability to trap heat more effectively than carbon dioxide. Global warming potential (GWP) is the globally averaged relative measure of the warming potential of a particular greenhouse gas. It is defined relative to a reference gas. Carbon dioxide (CO_2) was chosen as this reference gas. This means that carbon dioxide has a GWP of 1. Other common greenhouse gases related to energy use, production, and other university activities are methane, nitrous oxide, and HCFCs. Table 2.3 shows a list of heat-trapping gases and their GWP relative to carbon dioxide. Note that the 1996 and 2001 IPCC reports use slightly different numbers for GWP. Other sources may use GWP numbers that vary slightly from these.

At colleges and universities, these other climate-altering gases are emitted in varying quantities depending on the type of institution, activities, and types of equipment. For example at Tufts, other heat-trapping gas emissions come from our School of Veterinary Medicine in the form of methane from our dairy herd. Institutions with agriculture schools will have methane emissions from herds and from composting operations as well as nitrous oxide emissions from fertilizers. Tufts' dental school also

Table 2.3
Global warming potentials of heat-trapping gases

Gas	1996 IPCC GWP	2001 IPCC GWP
Carbon dioxide methane	21	23
Nitrous oxide	310	296
HFC-23	11,700	12,000
HFC-125	2,800	3,400
HFC-134a	1,300	1,300
HFC-143a	3,800	4,300
HFC-152a	140	120
HFC-227ea	2,900	3,500
HFC-236fa	6,300	9,400
Perfluoromethane (CF_4)	6,500	5,700
Perfluoroethane (C_2F_6)	9,200	11,900
Sulfur hexafluoride (SF_6)	23,900	22,200

Sources: Energy Information Administration, *Comparison of 100-Year GWP Estimates from the IPCC's Second (1996) and Third (2001) Assessment Reports*, http://www.eia.doe.gov/oiaf/1605/gg03rpt/gwp.html.

contributes to heat-trapping gas emissions through the release of nitrous oxide, or laughing gas, which is used in clinics as an anesthetic. Nitrous oxide is recovered by dental clinics; however, some releases to the environment still occur.

All institutions are likely to use refrigerants that are heat-trapping gases. While the release of these chemicals is regulated and should be avoided, releases do take place when air-conditioning equipment breaks or when unskilled or unlicensed technicians make repairs. Research laboratories may also use and release some of these other climate-altering gases. For example, research on semiconductor manufacture also may release extremely potent heat-trapping gases. Venting natural gas, propane, or methane either accidentally (through leaks) or intentionally also contributes to climate-altering gas emissions.

A comprehensive climate change action effort at a college or university will identify those gases that are used and released at the institution and include them in their climate change action plan. Since some of these chemicals and their releases are regulated as air pollutants, ozone-depleting chemicals, or hazardous materials, there are many reasons to address them. Furthermore, most of these gases are far more potent contributors to climate change than carbon dioxide and can be traced to a limited number of sources over which only a few people have control. The limited number of sources can make it relatively easy to reduce the emissions of these gases. In contrast, the university's generation of carbon-based heat-trapping gas emissions is diffused over many use points throughout the institution.

Summary

Climate change basics can be challenging to communicate in a way that motivates political decision makers to take prompt and effective action. We hope that the efforts of colleges and universities to take climate action will help leverage wider understanding and more effective responses, perhaps in collaboration with regional, state, and local efforts now underway.

In general, wealthy countries generate more heat-trapping gases than their less affluent counterparts, and the limited data available on colleges and universities suggests that the same general relationship holds true for academic institutions. In a sense this is good news: those that are more affluent have more resources for addressing the problem.

Carbon dioxide is the focal point of our climate action discussion because humans produce it most abundantly as we heat and cool our buildings, generate electricity, and operate motor vehicles. However, smaller quantities of several other climate-altering gases may be generated on college and university campuses and these may have much more effect on the climate per molecule.

Just as nations prepare inventories of climate-altering gases to establish their base case and inform decision making and action, colleges and universities can and should prepare an inventory. In the next chapter we discuss campus inventories and climate action goals.

3

The Campus Inventory and Climate Goals

In this chapter we discuss the linked topics of goals for climate action and campus inventories of climate-altering gases. A baseline inventory enumerates the sources of emissions associated with the institution and is the starting point for discussions on what actions can and should be taken. As the inventory is updated over time, it becomes the basis for tracking progress toward goals.

Emission Sources and the Emissions Inventory

The most common heat-trapping gas from college and university activities is carbon dioxide released largely as a product of the combustion of fossil fuels. Direct emissions are those that occur from activities owned wholly or in part by the university. These include the emissions resulting from the combustion of fossil fuels for heating buildings, heating hot water, and powering the university vehicle fleet.

Indirect greenhouse gas emissions are releases from sources not owned by the university but occur as a result of university activities. The major indirect emission releases associated with Tufts result from the purchases of electricity and steam generated by a third party. Other indirect sources include emissions resulting from the commuting population of staff and students to and from Tufts, from deliveries, and from university-related travel on trains, buses, and aircraft. Indirect emissions also include those associated with the construction or renovation of buildings and the emissions associated with all materials used and purchased by the university such as the life-cycle emissions associated with the production, transport, and final disposition (reuse, recycle, or disposal) of goods and waste products. These goods include furniture, paper, computers, books, and food, just to name a few. Electricity generation is the single largest indirect source of climate-altering gases at Tufts. Preliminary studies of these indirect emissions show that air travel associated with university business may be the second largest single source of our indirect emissions.[1]

A greenhouse gas inventory is an accounting of all the greenhouse gases generated by an organization. Typically the inventory will consist primarily of the gases that are created by energy production, but it also includes those that are related to transportation and can even include emissions from animals or food waste. While there are numerous inventory templates,[2] an inventory is always unique to the organization completing it.

An organization may choose to complete an inventory for a variety of reasons including:

- Determining the college or university's impact on climate change
- Assessing progress toward a goal
- Informing priorities for action
- Measuring results of specific actions
- Communicating information about climate change

An inventory is needed to establish a baseline because it is not possible to use air-monitoring equipment to measure all of the sources of campus emissions. We cannot measure directly the climate-altering gas emanating from each cow in our research herd and we cannot measure the carbon dioxide from each car used for commuting to campus. Many of the emission sources need to be calculated using emission factors. An inventory is an essential starting point for emission reduction efforts. Here is an analogy. If you think you need to lose some weight and you plan to take action to lose a few pounds, you start by weighing yourself. Conducting an inventory is equivalent to stepping on the scale and recording the result. The inventory is not an action plan. In the weight analogy, the action plan is equivalent to your diet and your exercise plan, and they come later.

Appendix C contains detailed information on conducting a campus emissions inventory. Included is information on emission factors to use for various campus activities so that an accurate baseline can be created.

Why an Inventory Is Important

An inventory enumerates the amounts and sources of emissions associated with an organization or a geopolitical unit. There are many reasons to embark on the inventory effort. First, it is a tool to assist in the systematic identification and recording of known and unknown sources of climate-altering gas emissions at an institution. An indirect benefit of conducting an inventory is the knowledge gained of the structure and operation of the institution, not only for heat-trapping gas emissions, but for other pollutants and environmental stressors as well.

Second, the inventory will provide a benchmark against which improvements can be quantified. Essential to justifying the commitment of resources (i.e., spending money) is an estimation of the quantities of emission reduced related to a specific effort—how much will the proposed action reduce emissions and what will it cost? Quantifying the effectiveness of actions that reduce energy and material use and that lead to reductions of emissions will assist in the justification of resource allocations (time and money).

Finally, the inventory will be a reference to communicate the most important, as well as the not-so-obvious, emission releases. Information gathered in the inventory will be used to generate charts and graphs that summarize the importance and status of the emission reduction effort at the institution. Further, the inventory will assist in identifying the aggregate impact of the many actions, small and large, that emit climate-altering gases. While the environmental impact of one lightbulb, one meal, or one photocopy is difficult to assess, the cumulative effect of everyday actions creates a substantial ecological footprint. For example, in 2004, Tufts University, a community of more than 8,000 graduate and undergraduate students, served 2 million meals, made 21 million photocopies, printed more than 40 million pages, consumed 90 million gallons of water and 64 million kilowatt-hours (kWh) of electricity, generated over 4,000 tons of solid waste, and released more than 24,000 metric tonnes of carbon equivalent (MTCE) of greenhouse gases. Each source of emissions, identified and articulated in comparable units, establishes the basis on which an institution can take action and eventually meet its commitment to reduce its contribution to global climate change.

Thus far, we have concentrated on the role of an inventory with respect to internal college or university decision making. There are many other potential roles, particularly as political jurisdictions and other interests begin to take action on climate change. Early efforts to create a market for carbon rely heavily on inventories that are prepared consistently. For example, the Chicago Climate Exchange has created emission reduction goals against an audited four-year baseline. Other efforts to trade carbon such as the Climate Neutral Network and the Carbon Trust rely on a more project-by-project accounting of climate-altering gases.

Interpreting and Using Inventory Results

The inventory makes it possible to develop a baseline, track trends, and in the future, to measure and communicate progress. Some uncertainty will always arise in the data collection process due to human error, lack of standard reporting across departments, and estimation to compensate for missing data. In these early days of climate action, the important elements are current sources of greenhouse gases and enough historical information to reveal trends. It is also essential to keep careful notes on assumptions made so that subsequent inventory updates can easily trace the origins of each data element.

The method outlined in appendix C for completing a climate-altering gas emissions inventory provides a relatively quick and inexpensive way to document an institution's emission sources. This broad-brush approach will be most useful in identifying emission trends and areas of growth, suggesting spheres of influence and activities where emission reductions and energy-efficiency increases can have the greatest impact. It will also aid in revealing whether an institution is becoming more energy intensive. The inventory data will allow an understanding of whether energy demand is increasing on a per student or per square foot basis, or both. However, an inventory only informs and does not, by itself, determine priorities for action.

Strategic Implications of the Inventory

The inventory and the trends it reveals give decision makers a sound basis for establishing a strategic direction. If trends reveal that emissions are increasing, the inventory may help identify which segments are of greatest concern. If the data indicate an upward trend in emissions associated with heating buildings, it will be important to first understand why this is occurring. Are increased emissions associated with the addition of new buildings to campus? Or are emissions up because there were several warm winters followed by a particularly harsh winter when an exceptionally large amount of fuel was used? These different explanations suggest very different actions on the part of the college or university. If emissions are increasing because more square feet of building are being added to campus, it is important to scrutinize construction standards to learn whether future buildings can be made more energy efficient as a design goal.

Trends in emissions may also be temporary and linked to fiscal rather than climate-driven decisions. If a college or university has dual fuel systems in which either oil or natural gas can be burned, the institution may follow an energy strategy to use the least costly fuel. This means that when the price of natural gas falls relative to oil, use of gas will reduce heat-trapping gas emissions. But if oil is less costly, the switch to its use will increase emissions.

Tufts' Inventory

The greenhouse gas emissions inventory for Tufts University shown in figure 3.1 indicates an upward trend in the emission release levels from 1990 to 2005 despite our efforts at reduction. These emission increases were on a universitywide basis and also on a per student and a per square foot basis. Comparing the inventory to the baseline and to past inventories reveals that our electricity is increasing due to plug loads from the escalating number of personal computers and other electrical equipment and the equipment-use patterns of students, faculty, and staff. This kind of increase is particularly difficult to reduce because behavior changes on the part of equipment users are required.

Tufts is becoming more energy intensive, both on net and by normalized metrics. This means that if trends continue and no further emission reductions are taken, Tufts will not meet the Kyoto goal. With record high electricity prices predicted for the future, the trend of increasing electricity use is not exclusively a climate-related concern. It means that the university will continue to face increasing operating costs. This convergence of financial interests and climate change interests is precisely the kind of opportunity that has the greatest likelihood of resulting in creative long-term solutions that satisfy multiple organizational goals. In addition, with very high electricity costs projected, the range of feasible

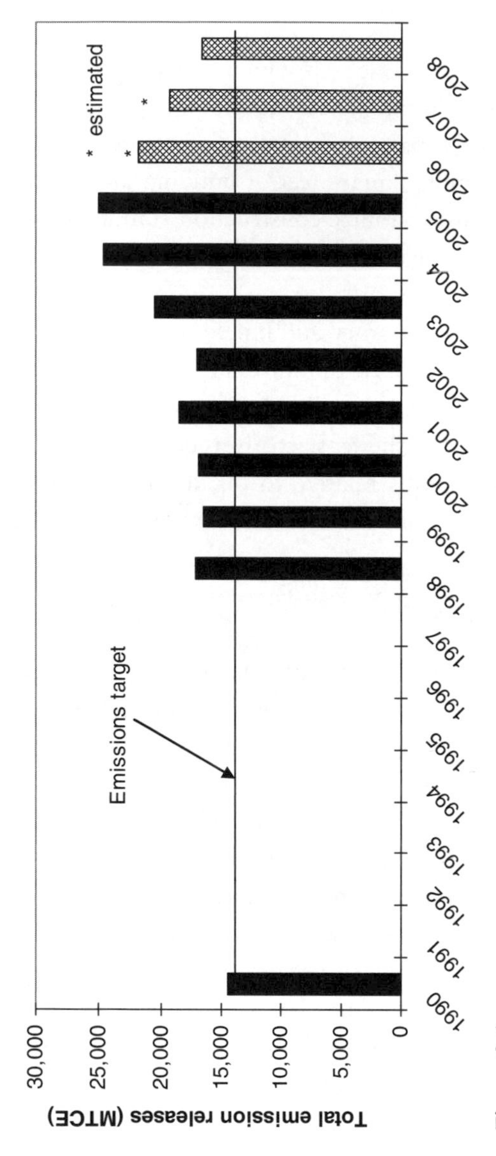

Figure 3.1
Tufts Carbon Inventory

options may be broader now than it was just a few years ago. As electricity costs increase, energy-efficiency projects pay back more quickly. Renewable energy, which usually is more costly than fossil-fuel-derived power, may also be more competitive than it was just a few years ago.

In January 2006, Tufts changed electricity providers. Because the new provider will increasingly use hydropower combined with cleaner burning natural gas, the university's emissions from electricity will decrease, as shown in figure 3.2. This change in providers made sense for the university both to reduce emissions and to achieve a measure of stability in energy costs. Energy providers that rely on alternatives such as hydro, wind, and solar do not experience the price volatility associated with fossil fuels.

Limitations to Our Inventory. The majority of the inventory data we collected at Tufts were related to direct emissions from fuel use, and indirect emissions one level upstream for electricity and steam energy purchases. In general, access to the fuel purchasing data allows for an inventory of the major sources of direct greenhouse gas emissions. Indirect emissions data, such as estimates of commuter transportation fuel use; materials purchasing, consumption, and end disposition; new construction; and facility renovations are more difficult to obtain. The difficulties are due in part to the lack of data as well as to the labor requirements and expense of an expanded boundary for data collection.

Data regarding the amount of fuel required to deliver a pound of steam were also not readily available, so we had to make an estimate. The quantity of steam generated or purchased is a substantial source of CO_2 emissions at Tufts. It is the second largest source after electricity. Our Boston campus purchases steam from TRIGEN, Inc. TRIGEN estimates a maximum of 90 percent efficiency based on the coproduction of three energy products: electricity, steam, and chilling water. For a better understanding of the amount of energy required to deliver a pound of steam, further information regarding the efficiency of the other two products is necessary. At this time, our inventory uses 30 percent efficiency for steam. If the system were 100 percent efficient, this would reduce Tufts' emissions by as much as 25 percent.

Prior to 2006, calculations of emissions based on electricity purchases also used averages for our utilities. If a utility engaged in fuel switching

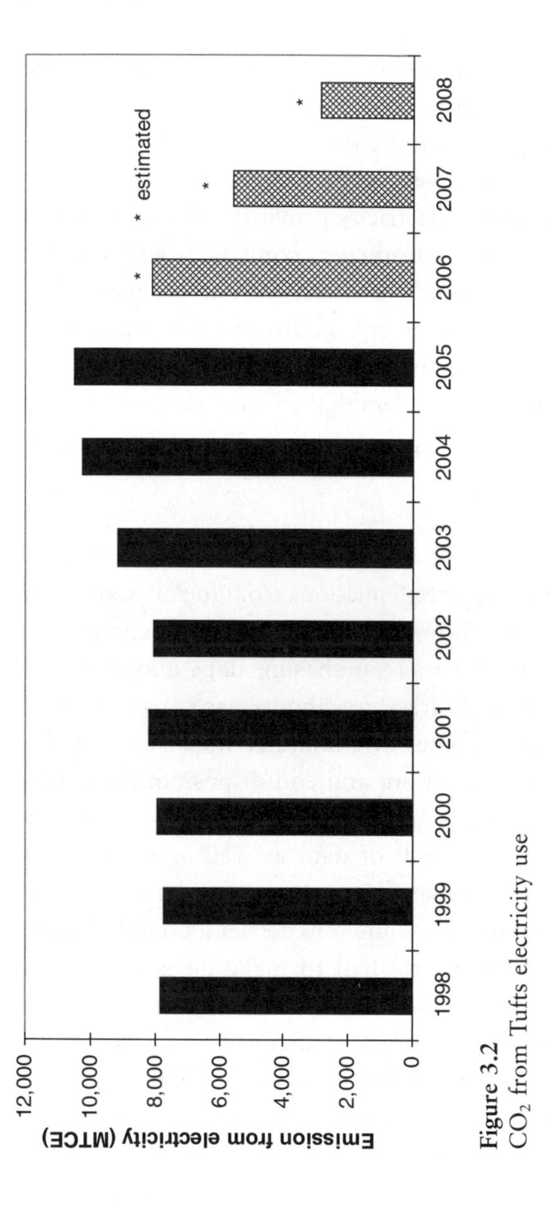

Figure 3.2
CO_2 from Tufts electricity use

in response to price or availability, not all of the emissions implications were captured in the intensity factor we used. This also meant that if one of our electric companies accommodated peak summer demand by bringing online a very carbon intensive power plant, the increased emissions from our use of this peak power probably were not reflected in the inventory. Because electricity purchases represent about half of Tufts' emissions, the average intensity factor could be a significant source of error in the calculation of total emission quantities. Figure 3.2 demonstrates the dramatic effect of a switch to a less carbon intensive fuel mix.

In the future, we hope to expand our emissions inventory to accurately include secondary emissions, including those associated with materials and equipment on campus. This will include the energy embodied in new construction and building renovations as well as the energy embodied in goods we purchase. This expansion of the inventory will capture a larger portion of the indirect emissions attributed to the purchasing choices by Tufts' staff, faculty, and students. By capturing the broadest boundary of influence by a university, appropriate choices for action will be better realized. But even a qualitative recognition of these indirect sources can be used in policy formulation and action plans.

Despite its limitations and its reliance on estimates for some quantities, the beauty of the emissions inventory is that the principles of the universitywide assessment can be applied to individual measures. For example, the same method for calculating emissions from aggregate electricity consumption can be used to calculate the avoided emissions from a single lighting retrofit that reduces energy use from electric lights. Or avoided emissions can be calculated from the installation of a photovoltaic array that would generate power from the sun's energy rather from fossil-fuel-burning power plants.

Perfection versus Utility of the Inventory—Striking a Balance

Conducting a climate-altering gas inventory for a large institution can be a complicated task. Some data are difficult to obtain, or may simply be unavailable. In these cases it may be necessary to use estimations. For example, you can estimate the emissions associated with commuting by making (and documenting) assumptions about average trip length and vehicle fuel efficiency. You can also estimate emissions associated with electricity by using use the statewide average emission coefficient published by the Energy Information Administration instead of using emission information specific to your utility. But because electricity is such a large source of emissions for most campuses, using utility-specific emission factors will give you better information. For example the Tufts Boston campus emissions decreased between 1990 and 1998, not due to conservation or the use of alternative fuels, but because the mix of fuels the utility used to generate power was changed to reduce coal and increase the use of natural gas.

Accounting for all of these factors can be time consuming. We caution those undertaking an emissions inventory to consider the time needed to perfect the inventory and compare it to the types of decisions being made based on the data. When we present our inventory to faculty who are familiar with the intricacies of heat-trapping gas inventories, we are often bombarded with questions about our methods or about which inventory method we favor (EPA, World Resources Institute, or others). And we are clear where we have made estimates, where we have good data, and about our inventory approach. However, the important message from the inventories of most colleges and universities, from our national inventory, and from most of our personal inventories, is that the trend over the last ten years has been a significant increase in emissions. This trend is independent of inventory method since each varies only slightly in the details. This upward trend must be reversed if we are to meet the challenges of addressing climate change and slowing its progress.

Goals for Action

How do you select a program goal for climate action? What are the implications of different goals? Tufts' commitment to the Kyoto goal was both a political statement and an institutional challenge. When we made the commitment in 1999, we did not examine alternative goals for climate action. Because several possibilities for alternative approaches have emerged among colleges and universities in recent years, we offer some thoughts on goal selection.

How ambitious should the climate action program be? Answering this question requires an understanding of the resources available (and potentially available) and the risk profiles of the institution and key decision makers. It may be useful or practical to have a short-term goal such as

Kyoto (7 percent reduction below 1990 level in 2012) or a more ambitious goal such as the one selected by the New England Governors and Eastern Canadian Premiers, whose short-term goal is 10 percent below 1990 levels by 2020 and whose long-term goal is a 75 percent reduction in emissions. Goals of this type, tied to external programs, can help in communicating both within the institution and with those outside. The time horizon is long from the point of view of student generations, but fits within the employment cycle of many university employees. In other words, in 2012, there will be people at Tufts who remember our making the commitment, and who are interested in knowing whether we met the goal we set for ourselves in 1999.

Another consideration is whether it is important to set a goal that the institution is likely to meet. In "selling" the Kyoto commitment at Tufts, top decision makers were told that there was a reasonable likelihood of meeting the goal. The importance of such a feasibility assessment will vary with individuals and organizations. It may be critical to getting "buy-in" for the concept; however, in organizations with different cultures the question may not even come up. Some colleges and universities may be attracted to the concept of setting a goal and may be unconcerned about the possibility of failing to meet the goal. More risk-averse organizations, by contrast, may limit themselves to commitments they are reasonably certain to meet. Hybrid approaches are also possible: companies often articulate stretch goals or aspirational goals to frame their environmental programs (e.g., zero waste) and establish a set of shorter-term targets that they can meet or beat, such as a 5 percent reduction per year over the next five years. Colleges can take a similar approach. But deciding not to commit to a goal or setting a goal that allows for growth in net emissions is another matter. If you understand the scientific basis for climate action, you also understand that continued growth in emissions lies at the heart of the problem. As we mentioned in chapter 2, the failure of an institution to take responsibility for reducing its own emissions puts a greater burden on others.

Innovative thinkers such as Amory Lovins argue convincingly that we need to take quantum leaps rather than incremental steps to address climate change. Lovins argues that actions *must* be definitive, swift, and comprehensive.[3] Yet we know that not all colleges and universities are in a position to take action of the type Lovins advocates, particularly if

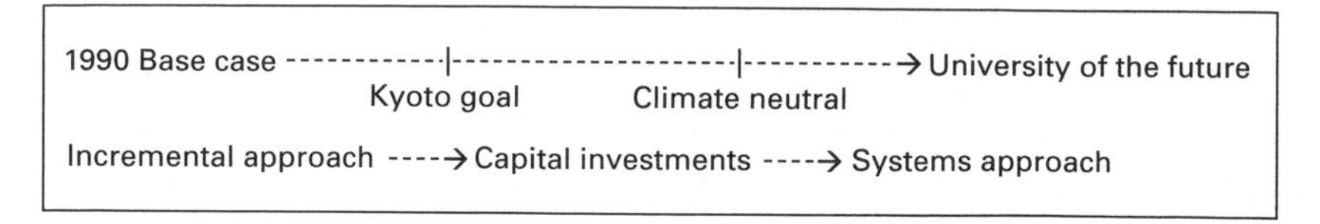

Figure 3.3
Continuum of climate action goals

the actions require significant capital investments or risk. Figure 3.3 is a continuum showing the 1990 base case and a progression to the emission-free university of the future. This hypothetical university of the future produces no climate-altering gases in the life cycle of its infrastructure development or in its operations. Moving toward the university of the future requires increasing amounts of systems thinking, risk taking, and capital investment. Our experience suggests that a great deal of emission reduction can be achieved with incremental approaches, in large part because many existing systems in colleges and universities are inefficient.

We expect to be able to meet Kyoto goals using a combination of incremental and systems approaches that we will describe in subsequent chapters. Many scenarios will allow us to meet the goals. For example, a combination of a 10 percent decrease in electrical use through efficiency, the purchase of 10 percent green power, and a fuel switch from oil to natural gas will achieve our goals. The final set of actions we will take will undoubtedly look different from the ones we now envision, but the point is that there are many ways we can meet the Kyoto goal.

Tufts has a modest endowment, and foundations and donors have not shown much interest in very large scale infrastructure investments needed for dramatic emission reductions, so a great deal of our progress in the short term will be achieved by reallocating funds. Fortunately, donor attitudes may be changing. Wellesley College made headlines in the spring of 2005 with the announcement that alumna Leonie Faroll made a record-breaking bequest of $27 million, the largest bequest to a women's college. Faroll specified that most of the gift will support maintenance and capital improvements at the college power plant, and the facility will be named after her parents, Berenice and Joseph Faroll. We thought the *Boston Globe* was a bit harsh when it described Faroll as "a frugal and eccentric woman."[4] We think terms such as trendsetter and innovator might be more appropriate to describe her action, and we hope that Tufts has an equally inspired and thoughtful benefactor. We remain optimistic that the day will come when donors to our institution understand the value of overhauling infrastructure systems and feel that naming rights to chillers, heating plants, and lighting control systems are as desirable as naming sports facilities and residence halls. But in the interim, we are grateful for the climate action support we receive.

Some colleges and universities may be able to embark now on a more systems-oriented and less incremental approach to emission reduction. Systems approaches may be consistent with an organization's decision-making culture, or may be desirable to meet multiple organizational goals. For example, the University of Massachusetts Boston campus previously used centralized electric heat for its buildings. When they switched to natural gas, university emissions dropped by almost 10,000 tons of CO_2 in fiscal year 2002. The project savings were estimated at $1.9 million annually.[5]

Even in an organizational context in which there is complete support of top management and adequate capital resources to pursue dramatic and comprehensive steps toward the goal of the university of the future, it is still critical to achieve buy-in at many levels in the organization, and to orchestrate a coordinated approach to achieving the goal.

Climate Change Goals Selected by Colleges and Universities

In these early days of climate action, institutions often express their goals relative to a 1990 baseline. Over one hundred college and university presidents in New England have signed on to the regional goals established by the New England Governors and Eastern Canadian Premiers. Table 3.1 summarizes goals at a variety of institutions, all representing commitments of their presidents.

Our experience suggests that efforts at campus greening or sustainability are strengthened by adopting a larger goal from outside of the university—a local, state, national, or international goal—as a way to focus the campaign and to develop a motivation for the institution to "do its share." We have found that the commitment to the international and regional climate change goals has several advantages over our earlier opportunistic campus-based greening efforts described in *Greening the* same heating system twice. With careful implementation, the number of double investments can be minimized, but it is a theoretical argument against the incremental approach. In contrast, a goal that clearly requires deep investments in systematic approaches, such as Oberlin's strategy to seek carbon neutrality, while more costly initially, may be more cost effective for the college and more effective for the planet over time. This book largely addresses the incremental strategies since these are the strategies that are more attainable by most institutions, but we also discuss strategies requiring large leaps from current business practices.

While we are attracted to very ambitious goals, we are also pragmatists. We are not willing to simply talk about lofty goals, even though ambitious goals are required to radically decarbonize our economy. We place value on taking actions now and learning from experience. In the short run, our institution's ability to act is limited by fiscal constraints and external policy impediments, so we will take whatever approaches we can until deep investments in systematic approaches become possible.

Monitoring Progress toward Goals

Monitoring progress is important for any program, and the inventory facilitates this process. The inventory contains the data that can be used in a range of decision-making and communication tools. Progress reports based on the inventory are valuable to the program manager and (in our case) the foundations supporting the Tufts Climate Initiative, as well as to the many people across campus who have taken action to reduce emissions. Equally important, progress reports can be a vehicle for engaging the attention and perhaps participation of people in a position to influence further resources and action.

Several challenges are associated with monitoring and reporting on climate action. The most obvious challenge is that despite efforts taken to reduce emissions, they may still increase. At Tufts, we saw electricity use per square foot increase about 3 percent per year from 1990 to 2000 despite implementation of the Green Lights program in 1991 and the efforts of the Operations Division to improve efficiency. Part of the net increase in electricity use is associated with the addition of more square feet of built space to the campus, but the increase on a per square foot basis is likely associated with increased "plug loads," or electricity-using equipment in student rooms, offices, and classrooms. (See chapter 10 for more on plug loads.) Between 1998 and 2004 the growth in electricity use slowed, but the addition of an energy-intensive laboratory building in 2005 reversed the trend. Metering allows us to identify patterns in some buildings; however, more investments in metering are needed for us to better understand energy-use patterns and develop tailored efficiency programs. In addition to metering, personnel time must be dedicated to analyze the data and identify the trends and technological solutions.

Summary

Colleges and universities generate climate-altering gas from a variety of sources and may be locally significant contributors to climate change. The inventory is a first step toward developing a climate action plan; it enumerates the sources and quantities of heat-trapping gases generated on campus. The inventory provides information to decision makers that can be used to develop a strategy for emission reduction, and it becomes a baseline against which progress toward goals can be measured. In the next chapter we examine campus decision makers and their interests related to climate action.

4

Climate Actors and Climate Advocacy

Who makes decisions that affect the campus emission profile? How can we convince these decision makers that climate action is important? On many campuses, it takes some form of political (with a small p) action to achieve large emission reductions because large reductions are often linked to large projects in which the stakes are high for many decision makers. And not all decision makers will place value on energy efficiency or reducing climate-altering gases. Here we talk about people on campus and the types of decisions they make that can influence emission reduction. On a few enlightened campuses, emission reduction already is a shared goal and all that remains is deciding what actions to take first; if this is the case on your campus, it makes sense to jump ahead to chapter 5. On most campuses, emission reduction is a passion of a few advocates who may not know how to transform their concern into a comprehensive institutional response.

Identifying emission reductions that are most likely to be consistent with priorities of the institution as a whole or with the priorities of individual decision makers is an important part of formulating strategy. It is also crucial to understand synergistic opportunities arising in the larger community outside the college or university. Thus the strategy combines information from the inventory (discussed in the previous chapter) with knowledge of the organization's fiscal and political climate, along with knowledge of the broader context in which the institution operates.

Understanding the Actors

A wide range of decisions have climate change implications at the university or college. Those with a direct impact on greenhouse gas policy that is compatible with trustee goals. The president (or top executive) generally serves at the pleasure of the trustees, and the trustees have final fiduciary responsibility for the institution. An important exception is offered by state colleges and universities, where legislatures establish budgets and assume responsibilities for deficits. In many respects, the function of college and university trustees is equivalent to a corporate board of directors in that they inform the strategic direction of the institution. These top administrators and trustees should be most attuned to the broad implications of climate change for university planning—both in terms of climate change action for energy planning and in terms of the impact of climate change on their campuses.

Deans are expected to implement policy directives, and serve as a critical liaison between the administrative and the academic functions of the university. Deans are in an ideal position to recognize the myriad ways that climate action can help achieve top management priorities and enrich the academic life of the institution. They can communicate to department chairs collectively and can also work with individual departments to identify projects or areas of interest. To the extent that deans have modest discretionary funds, they can indicate their availability to support climate change activity, whether it involves seed research, bringing a climate change speaker to campus, or a modest lighting experiment as part of a senior engineering design project. Deans can also insist on or support energy-efficient purchasing (see chapter 8) or policies (see chapter 9), as well as influencing new space and renovations on campus.

Most important, the top administration, as institutional leaders, can articulate the importance of climate change commitment and action and its link to programs; this in turn will influence decisions about facilities, construction, and operations.

Operations Department All of the decision makers on campus can influence climate action in their day-to-day decisions, but the most important allies for the climate action advocate are in the operational department of the college or university. At Tufts, the university's infrastructure is the responsibility of the Vice President of Operations. Although many functions are included in operations, those most relevant to this discussion include maintenance, construction, energy, food service, safety, and security. In recent years these responsibilities have seen an explosive growth in importance and complexity. Purchasing functions at many colleges and universities are often in this same organizational unit, although this is not the case at Tufts. Operations responsibilities intersect with climate change action in many ways that are complementary. The members of the physical plant or facilities staffs have very strong incentives to undertake activities that improve energy efficiency, promote water conservation, increase durability, and reduce operation and maintenance costs. For example, when energy-efficiency upgrades are undertaken and less energy is used, the university is less vulnerable to energy price spikes. Unexpected expenditures on energy may result in reduced funds for other operational services or for academic programs. This is a situation everyone wants to avoid.

At most institutions the facilities and construction groups are the only ones authorized to make changes to buildings and to enter into contracts with consultants or contractors to work on buildings, or to manage energy systems and energy contracts. This clearly makes sense, since the institution must have some oversight of its buildings for safety, aesthetics, maintenance, liability, and efficiency reasons. Often, however, we hear student, faculty, or other advocates ask "why can't we just install a *device* to save energy or reduce emissions?" While the device may indeed be a wonderful technology, its installation and maintenance must usually be coordinated with other existing systems that may be known only to facilities personnel. Without their knowledge, existing systems and the new device may fail. This fact of organizational life is one of the reasons we advocate establishing and maintaining excellent relationships with people who can facilitate device installation and ensure its proper maintenance and operation.

Facilities and energy staff, including managers, engineers, and technicians, are the first line of action on almost all emission reduction projects. These staff people remain extremely important even when a college or university relies heavily on outside consultants for information, design services, engineering, estimates, and implementation. In selecting consultants, putting a priority on designers and contractors who view energy issues as a priority and consider the energy implications of their project will reduce a university's operational costs over the long term and reduce carbon dioxide emissions. Communicating a commitment to such action is a key component of using this outside expertise to full advantage.

Outside of the university, decisions that affect energy reliability, price, and the availability of technology can have dramatic effects on climate change actions. The astute university will seize the opportunities presented by changing prices, power-purchasing cooperatives, power-generation opportunities, utility rebates, state or federal financing programs, renewable-energy promotions, and programs to promote energy efficiency. Planning for a cost-effective and reliable energy supply is extremely challenging, in part because deregulation has opened up many new options. Depending on faculty interest and expertise, this may be a fruitful area for faculty and staff collaboration.

Academic Departments Academic departments are responsible for managing within budgets, as well as for providing undergraduate and graduate educational experiences that conform to university standards and practices and meet relevant accreditation standards. Department chairs report to deans. Chairing practices may vary across departments and vary from one college to another. In some departments, eligible faculty serve as chair for a three-year term and then the responsibility automatically passes to another person. In other departments, chairs make an initial commitment of three years and if both the department and the individual continue to be satisfied with the arrangement, the person can remain as chair for an unlimited number of three-year terms. Other variations are possible, especially with respect to length of term. Chairing a department offers rewards and challenges in varying proportions. Faculty serving as chairs may or may not have significant managerial skills, and their interest in and commitment to the position also may vary.

Department chairs may spend considerable effort on personnel issues, making administrative staff decisions, mentoring junior faculty, and creating committees to search for new faculty members. In these activities there may be some opportunity to address climate change decisions, but the impacts are unlikely to be dramatic. For example, a chair might encourage carpooling or use of public transportation. The more substantial opportunities for a department chair to affect climate change decisions relate to curriculum and planning and operating the department's physical facilities. If a department is growing or undertaking new programs that require modified or new space, the chair may be in a position to advocate for renovations that minimize climate-altering gas emissions or for a new high-performance, low-energy building. When department chairs prepare or update their strategic plans, they can signal to the administration the importance of working toward physical facilities that are designed to minimize emissions. But chairs at some colleges will have to follow through with personal involvement and ensure that appropriate expertise is secured when the renovation or new construction is planned and executed. In some cases, chairs also can influence day-to-day decisions affecting climate change, particularly if there are no institutionwide policies regarding minimum and maximum temperatures, use of incandescent lighting, and purchase of energy-efficient office equipment.

Faculty The broad categories of faculty activity traditionally include teaching classes, advising students, conducting research, and engaging in service to the community. Depending on the individual and the institution, many of these activities may be linked by a theme and by an advocacy position. For example, a faculty member interested in climate change might teach courses on international environmental policy, energy and the environment, and environmental technology, and might conduct research on a range of technology and policy issues related to climate change. For service, a faculty member might serve as director of the university's climate initiative, and as faculty representative to the administration's energy affairs council. Such a person would naturally attract thesis students with similar interests, and through external speaking engagements and other activities in the larger community, the faculty member would develop contacts with individuals and organizations that students might pursue for internships and job opportunities.

Faculty members also participate in a range of professional research organizations related to their field or interests. For example, the Intergovernmental Panel on Climate Change, whose reports consolidate peer-reviewed work from around the world, is supported by a vast network of scientists and policymakers, many of whom are college and university faculty. In the United States, the National Research Council, the science and policy research organization of the National Academy of Sciences, relies on top people in their field, many of whom are university faculty, to advise government decision makers on issues such as climate change. See box 4.1 for select National Academy reports related to climate

motivated by a single individual or a small group. Occasionally these efforts fade as student leaders take a year abroad or graduate, so a challenge for advocates of climate action is to institutionalize programs that are valuable to the campus.

The primary vehicle for student engagement at Tufts is through coursework and internships in environment-related courses and programs. These course-related projects can be highly informative for students as well as useful for gathering background information about attitudes, technology, or policies, and we address them in chapter 11. We are also working to shape student action on climate change through a program called Eco-Reps, a concept begun at Dartmouth and modified at Tufts. Our Eco-Reps receive modest stipends in exchange for biweekly meetings and targeted efforts to carry a message of environmental stewardship back to resident students living in their dormitories. We discuss Eco-Reps in more detail in chapter 10.

Campus media and student government have a role in educating the university community, and are an important source of interactive communication. At Tufts, both student media and government tend to be reactive bodies, so climate change issues are covered in the campus paper when events are held, awards are given, or speakers are brought to campus, and they become the subject of debate in the student senate when petitions are launched. The *Tufts Daily*, a student paper, has featured several stories on issues related to climate action, often reflecting the interests of individual staff writers. One example appears in appendix F. Our experience is that students themselves are best at generating campus media attention for climate change, and their efforts can be exceptional.

Students can be excellent advocates for climate change. Often they either do not know or choose to ignore the unspoken rules for getting things done, a strategy that can produce positive results quickly. On the other hand, working outside the system also may alienate the very decision makers whose cooperation is critical to progress, or may lead to a single successful event rather than systemic change.

It is perfectly reasonable to think of students as the university's customers. Certainly many institutions think of prospective students this way as they compete for enrollments among the best and the brightest. Whether or not institutions treat matriculated students as customers (as in "the customer is always right") is another matter. But students are indisputably potential future donors, and that gives them a very different status from faculty and staff.

Students at several colleges, including Tufts, have also turned to the student body for climate action. Advocates have worked through student government on referenda to support the purchase of green power. At Connecticut College, students voted to add a modest fee to their bursar's bill to fund the green power.

Students also work in groups that reach across universities to promote sustainability and climate action. Examples include Kyoto Now!, started by students at Cornell[3]; StartingBloc, a student-founded organization focusing on corporate social responsibility and sustainability[4]; and the Campus Climate Challenge, whose goal is to engage 500 campuses in a long-term contest to reduce emissions.[5]

Engaging Climate Actors

We have found that climate change action, often spearheaded by TCI staff as advocates, is most effective when we give careful consideration to identifying the right decision maker and understanding the range of factors that influence their organizational role. Further, we try to bring in multiple players toward the same end. For example, we are always looking to bring student and faculty initiatives together with the appropriate facilities department so that efforts are realistic, forward looking, and pragmatic.

As we noted in chapter 1, colleges and universities are ideally positioned to take climate change action. They can establish modest experiments and evaluate the results, and when they do make significant investments, they can capture benefits over the long term. But changing business as usual can be extremely challenging, particularly if key decision makers are not convinced that a problem exists, or if it appears that change will cost money in an organization whose operating budgets are inadequate for the status quo. In this situation, a climate change advocate must identify areas of opportunity and key strategies for action. The business of a college or university "doing its share" to reduce greenhouse gas emissions may sound good in principle, but most decision makers will need to have the steps spelled out for them by the climate action advocate.

Organizing and Advocating for Climate Action

The climate action advocate has a challenging role to play, particularly in a large and complex organization. Although a great deal of effort has gone into understanding how change agents work in corporate environments, it is unclear how these insights translate in a college and university setting. And the climate change advocate is very certainly attempting to transform the planning and decision-making processes of many parts of the institution. In this section, we describe approaches taken and insights gained as we have advocated for climate change action at our campus.

The Art of Advocacy

Advocates or champions are often instrumental in advancing social and environmental causes, both in the public sphere and within organizations. For example, former U.S. President Jimmy Carter is associated with fair democratic elections around the world, and with affordable housing in the United States through Habitat for Humanity. Through the international nongovernmental organization Green Cross, former Soviet President Mikhail Gorbachev is associated with environmental protection. British Prime Minister Tony Blair is an increasingly vocal advocate for climate action.

Throughout this discussion we use the term *advocacy*, mindful of the fact that there are a wide range of cultures on college and university campuses, and the climate for advocacy is one factor that distinguishes institutions. We are also aware that the term *advocacy* can have negative connotations. We view advocacy as a noble undertaking, and hope that our use of this term is taken in a positive light.

Often advocates will be respected or widely recognized members of the community or will be charismatic, engaging, or nonthreatening, although it is important to point out that these characteristics do not describe all advocates. Indeed, some very effective advocates are abrasive and annoying, occasionally prompting speculation that high-level decision makers accede to advocates' wishes primarily to get them out of their offices. Other advocates are less well known and have achieved their goals because they believe in the program, they are "get it done" types, and they have made themselves indispensable to decision makers.

Advocates vary in their style and degree of effectiveness, but in general advocates are very effective in assembling resources in support of their cause. The resources in question can take a variety of forms, including votes on a policy matter, dollars for research or action, or students willing to show up for a public protest. A hallmark of effective advocacy is recognizing what approaches are best suited to achieve a particular goal, and timing their delivery so that a reasonable pace of progress is accomplished. Needless to say, there may be a diversity of views about what constitutes reasonable progress.

Like elected politicians whose success often rests on their ability to identify and articulate the interests of their constituents and to turn those interests into policy positions and votes, advocates must be adept at understanding the self-interest of the decision makers they are attempting to influence. Successful advocates also will frame messages in ways that elicit a specific desired reaction. While politicians seek a relatively narrow range of reactions (vote for me or my legislation and/or provide money for the next election campaign), advocates often seek to influence behaviors with broad social implications such as increasing faculty and student diversity, wearing seat belts, consuming alcohol responsibly, and in our case, reducing emission of climate-altering gases.

A Place in Decision Making

Difficult as the assignment may be, we take the view that the climate action advocate is on very solid ground. Energy and climate change have a "place" in a vast number of discussions and decisions that occur at colleges and universities. The reasons include conserving financial resources over the long term, increasing the reliability of the energy supply, decreasing the operational costs of facilities, demonstrating vision and leadership in stewardship of natural resources, and expanding graduates' knowledge and ability to function as informed citizens.

The climate change advocate faces several challenges. Because so many decisions with climate change implications are made on campus, one challenge is to decide how best to allocate scarce advocacy resources. Another challenge is to gain access to the decisions or decision makers with the greatest ability to influence greenhouse gas emissions. A third challenge is to address a range of decision makers to increase opportunities or points of entry. Students may be the easiest group to reach, but because their emissions will be reduced primarily by modifications in their behavior, their emissions may be least amenable to reduction. If we could get all students to turn off their computers at least six hours a day and not bring refrigerators to campus, the savings would add up, but changing all those individual actions will be tough. On the other hand, college and university trustees may be extremely difficult to reach, but if trustees were to make climate change an institutional priority and insist on the development of and strict adherence to ambitious policies related to energy efficiency in new and renovated buildings, and insist on energy-efficient heating plants for the campus, substantial emission reductions will be much easier to achieve.

Advocacy for Climate Action

There is no guaranteed effective strategy for organizing a campus effort to reduce emissions of greenhouse gases. This is because there is variation in academic institutions' structures, governance and accountabilities, and priorities. However, several compelling approaches have emerged as colleges and universities embrace a range of environmental issues.

Three strategic approaches seem to dominate. The first is to institutionalize environmental issues with permanent positions in the operations infrastructure, the second is to assemble high-level committees to call attention to the most pressing issues, and the third is to create an internal advocacy position or organization. None of these approaches excludes the others and none is necessarily more effective. Less formal approaches include individual faculty, students, or staff advocating for climate action on an issue of immediate concern such as green power purchasing.

Role of an Internal Advocate

Within an organization, an advocate can go well beyond the simple message and sound-bite approach used in the public sphere, and can become a catalyst for organizational change. A great deal has been written about change in corporations because an organization's ability to respond to internal and external signals is often associated with profitability. In literature on companies, the term *champion* often is used for an internal advocate for "external" issues such as environment, diversity, and social responsibility. Particularly with respect to the environment, corporate champions often have used the strategy of emphasizing that changes in favor of the environment can have collateral financial benefits. In an academic setting, advocates may not always be able to rely on financial arguments, and as a consequence may have to craft a range of messages for different constituencies within the community.

There are three fundamental reasons to advocate for climate action:

1. It is fiscally responsible.
2. It is the right thing to do as a member of society.
3. It is consistent with the mission of the institution.

These reasons are valid for all institutions, but different decision makers within the same university will value them differently, and individual institutions will also have different priorities.

TCI as an Internal Advocate

TCI plays the important role of advocating for climate change issues within the university. Staff and faculty associated with TCI act as advocates in university decisions that affect climate-altering gas emissions, and these decisions take place at many levels in the organization.

When TCI was first formed, we set the stage for advocacy efforts by meeting with as many key decision makers as possible. Through this effort, it was possible to introduce TCI and to learn, at an operational level, which people within the university had operational responsibility for the decisions we thought were most closely linked to emission reductions. We also gained a sense of the receptivity of different decision makers to the university's commitment to emission reductions.

We used those who were most receptive as the starting point for our advocacy efforts. Not surprisingly, many people were receptive, and some were already actively engaged in related activities. However, many people who were receptive were unable to be responsive. For example, we discovered that some decisions we thought (or hoped) were centralized were not. This meant that instead of having one receptive decision maker, we had vast numbers of decision makers. Other people who were receptive were unable to be responsive because of the overwhelming demands of their day-to-day workload. Over time, we have worked with many of these decision makers to enhance their resources so that more decisions can be made in favor of emission reductions.

Some people found it difficult to understand how emission reduction related to their responsibilities. In parts of the organization with many decision makers with substantial influence over emissions, we have tried to create opportunities for learning and career development. For example, TCI held a workshop on high-performance buildings primarily for people in the construction department. The relatively small Tufts audience created an opportunity for decision makers to meet and pose multiple questions of people with experience in designing and operating high-performance buildings, some of which are on college and university campuses. In addition, TCI has provided funds so that construction department staff can attend workshops external to Tufts to meet with their counterparts at other institutions and to learn more about green buildings and the coordinated design procedures that produce good buildings.

Part of TCI's mission is to create a sustainable mechanism within the university to continually consider climate action. For example, TCI worked with university facilities personnel, financial representatives, and the Vice President of Operations to free up funds to create an Energy Reserve to pay for efficiency projects on campus and create a mechanism to repay the fund from operating budgets. This sort of advocacy will help create ongoing emission reductions within the university.

One of the great challenges facing TCI is that a myriad of decisions influence heat-trapping gas emissions, and with limited resources we must "pick our battles" carefully. TCI consists of two part-time staff as well as two faculty and a series of student research assistants. TCI staff are primarily responsible for gathering information on pending decisions at the university that will influence climate change. This includes plans for new construction, major renovations, changes in energy and fuel supply, major changes in equipment, and modifications to procurement policies.

Given resource limitations, TCI then decides which projects will be the focus of advocacy efforts, primarily based on potential for reducing emissions, linkages with other priorities, and likelihood of success. Staff also monitor external events and opportunities related to climate change, including energy policy, cost and supply issues, conferences and workshops, prospects for financial support, activities of comparable organizations, and government initiatives. This external scanning allows TCI staff to act as agents for transfer of information to university decision makers whose choices may be influenced in favor of emission reductions. For example, TCI staff were able to identify a program in Medford, one of the cities in which our campus is located, to use funding from the Million Solar Roof Initiative to support the installation of two small residential-size photovoltaic panels on a university-owned house. The university's participation helped the city, and the renewable resource demonstration benefits the university.

Advocacy by TCI staff goes beyond information transfer to providing a range of resources. We support existing efforts in facilities, construction, dining, and other aspects of operations to enhance emission reduction opportunities. TCI staff have conducted research, hired technical experts, organized training sessions, and provided grant-writing labor and expertise. In many cases, we act as a bridge between the ideas and idealism of faculty and students and the pragmatic needs of operations personnel.

Faculty and staff associated with TCI create opportunities for people at Tufts to give speeches, participate in workshops, and otherwise enhance the visibility of the university's climate change efforts. Sometimes these opportunities are pursued by TCI, and sometimes we create a situation in which a key decision maker delivers the message. This strategy is directed at both internal and external audiences. When a top university official such as the president gives a speech related to climate change, several important exchanges occur. TCI faculty or staff often will provide talking points to help shape messages and ensure accuracy. This creates an opportunity to answer questions, to talk informally about successes and failures, and to provide updates on progress. After receiving appropriate clearances, TCI then can circulate the top official's remarks to others in the organization whose decisions may be positively influenced as a result. This is one approach available to shape a community's norms that is compatible with the resources available to TCI.

If we had to sum up our advocacy strategy at Tufts, it would be to work ourselves out of a job. Specifically, our goal is to create a situation in which all decision makers on campus take the generation of climate-altering gases into account in all of their decisions, and routinely create and take opportunities to reduce emissions. When that goal is reached, the Tufts Climate Initiative will have made the transition to Tufts' climate initiatives. Until then, we will continue to act as advocates. We have taken a hybrid approach, forming a high-level committee, acting as an internal advocate, and supporting the ad hoc efforts of students, faculty, and staff to pursue issues of concern. A short-term measure of our success is identifying members of the campus community who are willing to take ownership for projects that directly or indirectly reduce emissions.

External Influences

Now that climate change is emerging as a concern of the general public, there are an increasing number of opportunities to form alliances and exchange information. Despite the federal government's reluctance to engage in a mandatory program of reducing carbon dioxide emissions, the EPA is encouraging voluntary initiatives. In addition, several state, local, and nongovernmental organization efforts advocate climate change action, some of which are ambitious and mandatory. A variety of benefits may be associated with these external advocates for climate change. They may produce useful materials, provide access to technical resources, conduct research that informs efforts on campus, enhance credibility, generate public awareness, and provide funds for actions on campus.

On the other hand, it is also possible that partnering with external advocates will unintentionally divert attention and resources from the core activities associated with reducing the academic institution's emission of greenhouse gases. Tufts Climate Initiative relies on "soft money"—that is, funds from outside the university along with "hard money" university support. As we consider various options for grant and foundation support, we face one of the real dilemmas of our program: our progress is measured in reduced emissions. Activities such as writing this book are very important in transferring information, and we hope, motivating others to take actions to reduce emissions. But the book will not reduce emissions at Tufts, which is how we measure our performance. Indeed, it can be argued that time spent writing the book detracts from our ability to meet our goal. The same can be said about members of TCI speaking at outside conferences or other events. Because the TCI mission includes external outreach, we understood from the beginning that some activities would produce emission reductions and others would not. The challenge is in establishing a balance of activities and seeking funding from organizations that will support direct emission reductions on campus at the same time that they advance other organizational priorities and the wider goal of reducing emissions throughout society.

University Decision Making and Climate Action

Although we argue that colleges and universities can and should be leaders in climate change, we also have to state the obvious: the primary focus of the institution is its academic mission. We view the academic mission and the climate change challenge as complementary. Some institutions have a mission statement or a set of principles linked to sustainability that complements the goals of emission reduction. In such a case, the university mission statement can be an important tool for advocacy.

University Decision Points That Create Climate Action Opportunities

Many university decisions have climate change implications, although the implications may not be obvious to the primary decision makers. These are opportunities for the climate action advocate to explain the link and to help ensure that the decision shifts the institution in the direction of meaningful climate action. Here are a few examples.

Master Plan The master plan for a college or university articulates a vision and goals and creates an opportunity for the institution to express its vision through the future development of its physical facilities as well as its interaction with its host community and the environment. As part of the master plan, institutions have the opportunity to set goals related to sustainability. These goals can be expressed as commitments to energy efficiency, use of renewable energy, healthy, local, or "green" materials in new buildings and major renovations, commitment to public transportation, and a host of other factors that, particularly over the long run, have a significant influence on the campus emission profile. The link between the master plan and climate action is discussed in greater depth in chapter 9. The creation of the master plan provides an excellent opportunity for climate change advocates, because processes for master plan development are usually open and inclusive. In a good master plan process, climate action advocates will be invited to participate by virtue of being members of the community, whether they are students, faculty, staff, or residents of the host community.

Investment of Resources We have found that there are often significant opportunities to influence change at a time when investments are already being made. For example, when a small student residence was slated for life safety renovation and a cosmetic overhaul, TCI was able to take advantage of the fact that the house was "off-line" and unoccupied, and that some investment was already earmarked for the project. We used the existing renovation processes to target the house for additional energy investments, including lighting controls, boiler replacement, insulation, and a solar hot water system. Some of the emission reductions were gained by reallocating university funds, and some were gained by having the Tufts Climate Initiative pay for technology that was new to the place. Details of this project appear in chapter 5.

More importantly, the construction of new buildings or the major renovation of existing buildings requires capital and time investments. Designing these buildings to be as efficient as possible from the beginning offers important opportunities for successful emission projects.

Fiscal Crisis or Energy Price Increase Climate change action is, in part, about doing more with less. *Efficiency* and *waste reduction* are words with environmental and fiscal meaning, and astute advocates will find ways to make this link. A fiscal crisis is an opportunity to challenge the notion that energy costs cannot be tamed or that business as usual must be maintained. Actual or predicted energy price increases can also be a catalyst for action. Energy price increases are an opportunity for climate action advocates because they often result in much shorter payback periods for efficiency and renewable-energy projects.

To be sure, fiscal challenges can also hurt climate change action if the institution is unable to make investments that pay back, if key staff are cut, or if the institution makes major decisions to outsource without adequate protection in contracts. For example, a contract that specifies a least-cost solution may result in lower first cost but possibly higher life-cycle (energy) cost for a product or system. This would be the case with a low-initial-cost refrigeration unit whose annual energy cost is considerably higher than the annual energy cost of a more expensive unit that is more efficient.

Reorganization of Responsibilities When people think of organizational change they tend to focus on the redistribution of responsibilities and the attendant implications for programs and for local resources. At TCI, we consider the implications for emission reduction decisions. As an example, when purchasing is the responsibility of a central group, it means that the climate action advocate has a limited number of people to educate about climate-sensitive products. For instance, the university should only purchase desktop computers that are energy efficient (e.g., listed as "Energy Star" products). If a decision is made to decentralize purchasing, the climate advocate has a much more difficult task because there will be people with purchasing authority in each department, project, laboratory, or program.

Negotiation of New Contracts Decisions that offer ripe opportunities for emission reductions include new contracts for services such as fuels, electricity, transportation, vending, or catering, or equipment such as telecommunications, copy machines, vehicles, or air conditioners. Each of these contracts can be improved by including attention to fuel type, energy efficiency, and emissions over the life cycle of the goods or services being provided. See chapter 8 for more ideas.

Permanent Positions Creating a permanent position to address an issue of concern is a common approach to signaling organizational priority. As the campus greening movement has matured, several colleges and universities have created full-time positions with the titles such as "director of sustainability" or "sustainability coordinator." In many cases, climate action is part of the sustainability coordinator's portfolio. When located in the operations infrastructure, the individual in this position will typically have other responsibilities beyond climate change, including other action-oriented programs such as recycling and carpooling. At Tufts, the sustainability coordinator is outside of the regulation-driven Environment, Health, and Safety Department.

Having a permanent sustainability coordinator position is increasingly an indicator that a college or university takes these issues seriously. Learning about the institution's staffing and meeting people with climate action responsibility is a great way to get started on climate advocacy. If a college does not have a permanent position with climate responsibility, advocates can learn who is responsible for sustainability and find out whether climate action is on the person's priority list. At some colleges, advocating for a permanent sustainability position will help institutionalize the commitment to climate action.

Summary

By understanding the types of actors on campus and the decisions they are likely engaged in, the climate action advocate can decide how best to approach buy-in for emission reduction if it is not a top institutional priority. Activities such as increasing efficiency and reducing electricity costs are consistent with the goals of many campus decision makers, and the climate advocate may play a very valuable role in helping identify opportunities for achieving both fiscal and climate goals.

5

Strategy and Tactics for Climate Action

What kinds of actions will reduce heat-trapping gas emissions on campus? What projects should climate advocates take on first? How can advocates use knowledge about the interests of different actors to achieve buy-in for emission reduction? Here we talk about different types of projects that can be taken, and discuss the trade-offs between the glamorous projects that increase program visibility and the hidden treasures that are underground, behind walls, in basements, and otherwise out of sight and ignored, but that can have high potential for emission reduction.

Types of Climate Action

College or university emission of climate-altering gases can be reduced in several ways. Most of these measures involve one of four types of action: (1) increased efficiency, (2) increased use of alternative fuels or green power, (3) fuel switching, or (4) reduced demand through changed behavior and expectations. Technology, policy mandates, social marketing, and incentives are all tools that can be used to achieve these results. Carbon trading or offsets are options for action; however, these will not necessarily reduce a college or university's emissions. Along similar lines, actions to adapt to climate change are important, but many adaptations will not reduce emissions.

Increased Efficiency

The most direct way to avoid the emissions associated with the production of a unit of energy is to avoid producing that unit of energy in the first place. Some efficiency measures are cost-effective and quick ways to reduce greenhouse emissions. An example of an efficiency measure is to replace incandescent lightbulbs in desk lamps used by faculty and staff with compact fluorescent light bulbs. The more efficient compact fluorescent bulb produces the same amount of light (measured in lumens) using less energy. For example, a 100-watt-equivalent compact fluorescent bulb will use only 23 watts.[1] Another example of increased efficiency is through cogeneration of steam and electricity; however, this is a more complex measure requiring considerable capital investment. Cogeneration utilizes energy that would be wasted to create other useful energy products. For instance, many universities create steam on campus for heating and cooling purposes; cogeneration uses part of the heat to generate electricity and recovers much of the waste heat as steam, which can double the energy recovery of the operation. While this idea is not new, the technology now exists to make it practical even for smaller users. An extensive range of innovative technologies for efficiency are also available to those organizations willing to seek them out. More detail on efficiency is in chapters 6 and 7.

Increased Use of Alternative Fuels or Green Power

The electric industry, which remained basically unchanged for a century, has undergone something of a revolution in the past decade. Many states have now deregulated their electric industries. What this means for colleges and universities or other purchasers of electricity is that there is now a choice of electric generators. With this choice, a college or university may now purchase some or all of its electricity from renewable sources. Renewable energy offers low or no climate-altering emissions (in the case of wind and solar), which makes it very attractive to an institution that is interested in reducing its heat-trapping gas emissions. In addition to power purchases, colleges and universities may install green power on site, by tying solar or wind to existing or new building systems. More information on green power and alternative fuels is in chapter 8.

Fuel Switching

Fuel switching refers to the shift from one fuel source to another, less greenhouse gas–intensive fuel. An example would be switching from an oil-fired boiler to a natural gas–fired boiler. This change helps reduce emissions and can often save the university money because of the increased efficiency of the new system.

Reduced Demand Through Changed Behavior and Expectations
Colleges and universities can reduce demand for energy use by encouraging personal actions—such as turning off computers when they are not in use—or by implementing policies for heating and cooling. If less energy is used, fewer emissions of heat-trapping gases will be generated. In chapter 10 we address personal action in detail.

Other Actions
Two categories of action that will not necessarily reduce an institution's direct emissions are carbon trading and adaptation.

Carbon Trading This is a market-based approach to carbon emission reductions. The market, through contractual agreements with its members, sets an emission reduction goal. The members then have the option of either reducing the emissions in their own facilities or purchasing emission credits through the market; these credits represent emission reductions that other members have achieved (above and beyond their own commitments) and are willing to sell on the market. The world's first carbon trading platform, the Chicago Climate Exchange, began trading in 2003; Tufts is a founding member of the exchange.

Adaptation This term refers to taking action to anticipate or minimize the impact of climate change. For example, in an area that anticipates increased rainfall and greater frequency of extreme storms, a college or university may adapt to changing conditions by increasing the capacity of its stormwater management system to reduce the likelihood of damage associated with flooding. Such an adaptation, while important in protecting university property from the effects of climate change, will not in itself reduce climate-altering emissions.

Tactics for Climate Action Advocates

Among the major approaches to reducing emissions listed above, decisions need to be made about how to begin and what to do. The strategy is the overall plan, and the tactics are the techniques used to implement the strategy. Climate action can involve influencing decisions at many levels of the organization, so it is difficult to provide a comprehensive "how to" list. Some tactics that have worked for us are articulated here, and if they are not immediately useful, we hope that they will spark the kind of thinking that generates institution-specific solutions.

Understand the Process

Understanding the process is a critical step for many types of advocacy, and this understanding has benefited climate change efforts at Tufts. By *process* we mean the ways the organization makes rules and decisions that affect the institution's emission profile. The process(es) will vary, because different campuses have different entry points for decisions that affect climate-altering gas emissions. Is new construction planned or in the works? How frequently are old buildings renovated? If the organization is undergoing a period of rapid growth and development, understanding the master planning process is a good first step. Many colleges and universities have master plans that are displayed proudly as adjuncts to their capital campaigns, and the process of updating the plans may be inclusive and public. If new construction is a priority on campus, the process for construction decisions will be a key opportunity for climate action advocates to raise awareness of opportunities for emission reductions.

If the campus is mature and/or funds are unavailable for new construction, climate action advocates will benefit from learning about the decision-making process related to routine maintenance as well as to larger maintenance projects, often called deferred maintenance. Like new construction, scheduled maintenance is an ideal time for climate action advocates to leverage emission reductions. If a boiler is going to be replaced, a superefficient model can be selected. The same is true of appliances and lighting.

Processes for purchasing, orienting new employees, waste segregation, and recycling all have emission implications. Some of these processes will be more fully developed than others, creating different types of opportunities. For example, on a campus without a sophisticated or comprehensive recycling program, recycling may be a good place to start building credibility for subsequent climate change advocacy.

Link Climate Change Action to Existing Projects and Priorities

Like businesses, colleges and universities operate in a competitive environment. This is obvious when issues are the quality of students, the reputations of the faculty, the volume of research dollars, the success of the capital campaign, and the size of the endowment. A great deal of strategic thinking is focused on these attributes of a university's profile. In general, the operations function is simply responsive to these strategic initiatives. If a decision is taken to admit fifty additional first-year students, the operations department makes necessary modifications to residence halls. If a donor promises funds for a new building for the music department, the construction department works with architects, engineers, and the department chair to complete the project on time, within budget, and to the satisfaction of the donor, the administration, and music faculty and students. Accomplishing these activities can be so challenging on a day-to-day basis that opportunities for high-efficiency buildings or green power may be missed. And herein lies a significant opportunity.

Also think about ways to link project components across departments. Some climate change projects will have benefits for one department (for instance, the department paying the utility bills) and costs to another (for instance, the department paying for the equipment). This is often true of efficiency projects. If you can push the decision up to a higher level, the link can be made that the institution as a whole benefits.

Build on Intellectual Interests of Staff

About a semester after the Tufts Climate Initiative was launched, we sat down with the Tufts Vice President of Operations for an informal review. He could have complained that we were conducting a carefully orchestrated campaign of harassment. We wanted more energy-efficient measures in a new building, we wanted to reduce emissions in routine renovations, we wanted to add photovoltaic panels to a graduate residence within two weeks to qualify for a special program, and the list went on and on. Instead of asking us to back off, he thanked us. He explained that our advocacy of emission reductions was providing an intellectual rationale for many otherwise disparate and responsive activities, and that it was exciting for operations staff to be actively engaged with a high-profile program.

Climate action can also build on the ideas and idealism of faculty and students. The climate advocate can bring together people from different parts of the organization, each of whom has some interest in a problem.

Identify Allies and Respect Those Relationships

Developing a strategy to advance climate change action on campus can be a subtle exercise blending knowledge, persuasion, and good fortune. A great deal of what TCI is trying to do involves convincing people, primarily within the university, to spend "their" money to "our" advantage. This is hardly a novel undertaking, since advertising agencies exist to convince us to spend our money on their clients' products and services. But the enterprise assumes a different dimension when the metric against which decisions are measured is reductions of climate-altering gas emissions.

One of the first activities pursued by TCI was the development of strategic alliances or partnerships. Good working relationships within the university operations department were a higher priority at the outset than broad faculty or student involvement. An extremely important partnership is with the university's energy manager. The energy manager has the data needed to craft the emissions inventory along with an understanding of institutional history on energy-related issues, and knowledge of policies that are existing, emerging, and needed.

Working with the energy manager also illustrates the multiple challenges of creating learning opportunities for students and making progress toward goals. In an ideal world, many more projects would involve students working directly with the energy manager, or manipulating energy-related data. At Tufts, the energy program is short of staff, so we try to be very respectful of time demands. For example, we feel a responsibility to protect the energy manager from having to provide data to students or to review or supervise their projects. One of the dilemmas associated with student projects is that they can represent a significant time sink for some university decision makers, especially when projects are nonspecific or students lack knowledge of the "real world." Our experience suggests that typically the most valuable projects emerge from situations in which there has been the greatest level of interaction with the decision maker. Student projects are examined in greater detail in chapter 11.

Given the set of actors at the university, we have pursued multiple tactics in advocating for emission reductions. Each time we develop a concept or activity, we consider who might be our natural allies, and where the locus of power lies. For example, if our interest is having a rough feasibility assessment of the solar potential for selected buildings, we go to an engineering faculty member to find out whether a suitable project can be crafted for a senior design course or whether someone can engage a student in the undergraduate summer scholars program. Faculty have both the experience and the authority to make decisions about projects for their courses, and can sometimes create research opportunities with discretionary funds.

When TCI was making arrangements to acquire its first hybrid vehicle, we thought carefully about who might be a strategic driver. We considered top administration and security for their symbolic value and visibility. We finally settled on the grounds manager on our main campus in Medford. The grounds manager is literally all over campus every day and previously drove a gas-guzzling Crown Victoria, which was retired from the university security fleet when the hybrid Prius went online. TCI gave the Prius a distinctive paint job and this manager has become a key ally.

Consider Explicit and Implicit Incentives

Some problems linked to climate action require complex solutions. For example, our academic departments have no direct fiscal incentives to conserve energy. This is because the heating, cooling, and electricity bills are paid from a central overhead account and not from department budgets. Because some of the buildings at Tufts are old and perform poorly, in the winter it is possible to see people in one part of a building with space heaters blasting away, while those in another part of the building have open windows. The department is responsible only for the cost of the heater, so purchasing a space heater for an employee who is cold is a sensible decision under the current system.

If the department chair were to request a systemic fix to the building's heating problems, the department might be asked to pay for a share of the labor and equipment, and surely would have to accommodate the disruptions that might be associated with any major renovation. Clearly it is less costly and less disruptive for a department to resort to the short-term fix of using a space heater, although the costs to the university and the implications for the environment are significant. The existence of perverse incentives, along with the need for high-level attention to energy issues, led TCI to advocate for the formation of an Energy Affairs Council composed primarily of financial officers from the different schools. Although we do not expect rapid resolution of long-standing systemic problems, as a general strategy we are hoping to increase accountability for energy use.

In this case, one solution is to make individual departments responsible for their energy costs. The approach of holding smaller units of the organization responsible for what previously were group overhead costs has been used by large companies seeking to reduce impacts on the environment, and it has resulted in cost savings and waste reduction. While there may be immediate practical barriers to colleges and universities taking such an approach, such as the absence of departmental energy metering, it is a direct approach worth considering, particularly as energy costs rise. With such a shift in responsibility for energy costs, cascade effects can be anticipated, including new pressures for priority positions on the deferred maintenance list.

Another idea being practiced by companies serious about emission reductions is tying bonuses for select personnel to documented energy reduction. SUNY Buffalo has implemented some environment-related goals in its performance management system. This approach has challenges, not least of which is identifying an equitable compensation scheme, but the accountability gained is very valuable as we transition to the university of the future.

Work with Related Issues

Sometimes it is strategically important to explain the relationship between climate change goals and other campus initiatives such as recycling or environment, health, and safety. In addition it may be strategic to establish a link with larger university priorities such as regulatory compliance, cost containment, or religion.

Recycling An effective recycling program is a minimum standard for an institution that takes environmental issues seriously. If recycling containers are not available or if there are other concerns, we hear comments. When we hear about it, it is sometimes linked to issues of energy and commitment; "why should I turn down the heat when the custodian just mixed my recycling with the trash?"

How much can usefully be said about the relationship between climate change and recycling depends to an extent on the audience. A great deal has been written about attributes of recycling, and some assessments raise questions about net gains or benefits.[2] As with many complex questions related to the environment, a definitive answer may not emerge until there is agreement on where to draw the analytic boundaries. Recycling of aluminum is a case of a net benefit. The entire chain of events from raw material extraction through to manufacture of a product is extremely energy intensive. Recycling products such as aluminum beverage cans consumes considerably less energy, and thus creates fewer carbon emissions, than new-can manufacture.

Environment, Health, and Safety The environment, health, and safety (EHS) program at an institution has potential synergistic relationships with climate change. Typically, EHS personnel spend a great deal of their time dealing with aspects of the university's operations that are regulated by federal, state, and local authorities. Although not exhaustive, the list generally includes air pollution permits, wastewater discharges, hazardous and solid waste transport and disposal, low-level radioactive waste, pathology waste, and indoor air quality. Some of these programs have a direct or indirect link to climate change.

One area of direct climate change impact is the use of fume hoods, an energy-intensive technology that is necessary for healthy and safe use of chemicals. Fume hoods are essential equipment in laboratories, but their design, maintenance, and technology can have dramatic energy implications since the air that is exhausted from the system must be replaced by newly conditioned "makeup" air to the building. For example, at the University of Massachusetts at Amherst, the science center retrofitted fume hoods with automated controls to reduce airflow volumes through the hood during periods when they were not in use or when the hood's sash (door) was pulled down. At Tufts, a student project demonstrated that in some cases energy savings accrue with low-volume, constant-volume hoods.

Often the climate change link is indirect for the environment, health, and safety personnel. Nonetheless, we have come to recognize that Tufts EHS employees are routinely deployed throughout the campus and can become the eyes and ears for energy-saving opportunities as well. In addition, Tufts EHS personnel meet with all new employees, new teaching assistants, and new faculty using chemicals and provide them with basic orientation and training. We have worked with EHS personnel to include our brochures asking members of the community to turn off computers and lights as part of their standard message.

Regulatory Compliance As federal and state air quality regulations have become more stringent, university heating plants have come under increasing scrutiny. Institutions with relatively old heating plants may face difficult decisions that benefit from carefully done broad assessments. Old plants are a great deal less efficient and more polluting than models now on the market, but in some cases, old plants can be upgraded with fewer procedural hurdles than would occur with the construction of new plants. University decision makers may be reluctant to replace an old unit knowing that it will require an exhaustive and expensive regulatory process on top of the capital cost of the new equipment.

Tufts faces this dilemma on its Medford campus. We could likely meet the university's Kyoto goal by simply taking one action: replacing the central heating plant with a more efficient model, especially one that generates electricity and steam. We believe that in addition to environmental benefits, there would be cost savings and other positive outcomes. At the same time we understand that for a variety of reasons, including regulatory and other technical hurdles, a new heating plant is not currently a high priority of top decision makers. We know we cannot count on a new plant for emission reduction in the short term, but we keep informed of discussions related to the central plant, and remain optimistic that in the ever-changing calculus of benefits and costs, a new central plant will acquire increasing strategic value for influential university decision makers.

Cost Containment For many academic institutions cost containment has long been an issue, but it has gained increasing importance as fuel prices have risen and as federal and state priorities for education spending have shifted to the primary and secondary level. Reducing emission of climate-altering gas and cost containment are particularly compatible when the focus is on reducing electricity use. Historically, electricity use on campuses was predictable, and at Tufts, annual costs were established by regulated utility rates. With deregulation, energy costs have become considerably more volatile, and at the same time, electricity use on campus continues to increase. These factors, along with new concerns about the reliability of energy supplies, may create a forum for considering actions such as distributed generation and investments in renewable energy that an institution likely would not undertake solely in pursuit of a climate change goal. Price spikes in 2005–2006 prompted reconsideration of energy issues on most campuses.

Religion Environment and global warming are becoming focal points for action in congregations around the world. On campuses with active religious groups, it may be particularly effective to link personal actions with the values and beliefs articulated by religious leaders. In addition, several denominations have prepared action-oriented materials for their congregations that may help to set priorities for or otherwise inform campus action. And interdenominational groups are taking action as well. For example, Interfaith Works, formerly Interfaith Power and Light, was established as a direct religious response to climate change and has chapters in states across the country.

Determining Strategy

The climate action strategy is the overall plan for achieving emission reduction goals, and the plan is composed of a set of projects. Each college or university will develop its own set of criteria for selecting projects and setting priorities for climate action projects. Some questions that inform our strategic decisions include: How much will it cost? Is there a source of funds? Is this project on someone else's priority list? How much emission reduction will occur? How much effort will it take? Who can help? Will we make headlines for taking this action? The "right" mix of actions depends on a detailed understanding of the organization and the extent to which climate action is or can become consistent with other organizational priorities.

Selecting Projects: Going for Guts or Glory?

Climate action projects can be splashy and grab headlines (the glory part), or they can be obscure changes to mechanical and electrical systems (the guts). In a complex organization such as a college or university, the mission statement and goals are a useful point of departure in understanding institutional priorities, but they are just a start. Organizations educate their students in a myriad of ways, and finding approaches that embrace climate action is a good point of departure. Finding compatibilities with the less clearly articulated priorities of the organization may be more challenging, and in some circumstances, may lead to more substantial reductions in emissions. We take the view that all projects related to climate action are useful, but we place priority on understanding what emission reductions will be achieved and what will be the value to the organization of different climate actions. Table 5.1 shows the relative climate change benefit from a range of projects. This information may be useful in crafting an overall strategy.

Large-Scale Projects

Large-scale projects such as replacing or upgrading the central heating plant or constructing a high-performance building to replace an existing structure will make an appreciable difference in campus emissions but will require careful planning, long time scales, and both capital and human resources. Most large-scale projects that create opportunities for significant emission reductions will be undertaken because the projects address multiple priorities. For example, new space for life sciences

Table 5.1
Climate change benefit of specific projects

Project	Carbon dioxide reductions (annual tons)
Small photovoltaic (500 W)	0.3
Small solar hot water	10
Medium photovoltaic (23 kW)	13
VendingMisers campuswide	150
Steam traps campuswide	500
Occupancy sensors campuswide	900

research may be built to attract and retain top-quality faculty and students and externally funded research. Additional undergraduate residential space may be added to ameliorate concerns about encroachment in the surrounding neighborhood and to foster a greater sense of community among students. In both these examples, the climate change advocate may be successful in pressing for a high-performance building, or even a building that generates no net emissions; however, it is important to understand that the climate change aspects of the new construction are an "added benefit" of the priority agenda. Climate change may be used as a "hook" to attract financial support for the new building and the college may capture the educational and public relations benefits of the building's emission reducing features. However, climate action will rarely be the central motivating force behind a large-scale project.

Large-scale projects such as new buildings are usually distinguished by high-level team efforts that include trustees, top members of the administration, facilities and construction personnel, architects, engineers, development efforts, donors, and liberal doses of money. The climate change advocate may be successful in influencing the way all of the involved people understand the building, and may be successful in leveraging decisions; however, the climate change advocate will not generally have the final word in all of the decisions that influence the building's emission profile. Such projects are generally "too large" to let a single priority or individual dominate in most campus decision-making structures. The exception is an institution that is fully attuned to the myriad values of emission reduction; in that case, climate goals and other building goals may be intertwined.

Pilot Projects

Pilot projects are an opportunity to test ideas or technology when outcomes can be monitored and risks can be managed. For example, on a campus where students have no experience with front loading clothes washers, a pilot project might be done in a single residence. Energy and water use can be monitored, and students can be surveyed more than once during the pilot period to learn their attitudes about the washers and to understand whether measures might be taken to facilitate implementation. For instance, a letter might be sent out in advance of washer replacement explaining why the change is being made, what benefits will accrue (e.g., fewer emissions for the university and less drying time for the students), and what procedures students will need to follow with the new equipment (use less detergent—another savings opportunity). In addition to surveys, pilot projects frequently gather valuable information through informal feedback mechanisms such as comment sheets and selective interviews.

Pilot projects are particularly valuable on campuses because they frequently can be conducted by students and yield valuable results in a year or even a semester. With technology that is new or new to the place, a pilot project can be an invaluable source of comfort to a decision maker who thinks the idea is interesting, but has concerns about how well the technology will be received. The experimental nature of the pilot is consistent with the learning culture of educational institutions, and the small scale means that if the project is a failure at least it will be a small one. This helps save face for project advocates and minimizes expenditure of resources. If students are involved in a pilot, a practical failure can still be an academic success. Term projects and theses can document and analyze a wide range of failures, and yet they can result in exemplary course grades or degrees for their authors and valuable insight for future efforts.

High-Visibility, "Glamorous" Projects

Glamorous projects are those that come to mind when people think of emission reductions and climate change. In this category we include such efforts as small- or large-scale solar projects, purchasing green power for a single building, or placing a wind turbine or two on a visible hilltop or other prominent campus location. These projects are wonderful because they may attract positive press both in the campus community and more widely, and because they are an opportunity to raise awareness that people on campus are taking climate change action. Table 5.1 shows how these projects have contributed at Tufts.

Glamorous projects may be eligible for funding through government programs, utilities (often in the form of subsidies or rebates), or grants from foundations. Because glamorous projects are often small scale, they may require relatively few steps for approval and implementation. For some glamorous projects, a champion may be willing to take on responsibility for ensuring that the project is successfully launched. For example, a student or group of students may find a champion, secure approvals, and conduct the research needed for small-scale green power purchase. Finding students to monitor glamorous projects may be relatively easy, so collecting and analyzing effectiveness data may be less challenging than for other emission reduction activities. On the other hand, some of these glamorous projects result in modest reductions in emissions. It may be tempting to place priority on glamorous projects because of the attention they attract and because of their educational value; however, it may be possible to gain greater reductions in emissions by making considerably less glamorous efforts. In our view, it is easy to confuse glamour with guts—the projects with high potential for emission reduction.

Hidden Treasures—Low-Visibility Projects with High Potential for Climate Gas Reduction

Hidden treasures are projects that few people associate with climate change action. In fact, few people ever think about these projects at all. Many are truly hidden, out of sight behind walls, in basements, or underground. In this category we include steam traps, occupancy sensors, efficient motors, appropriately sized heating, ventilating, and air-conditioning (HVAC) systems, and building commissioning. If applied to existing systems or new construction, these measures can yield substantial reductions in emissions. However, these projects can be complex and challenging to execute even though they will yield energy (and thus monetary) savings throughout the project life.

The challenging reality is that for many institutions very substantial reductions in emissions will come from conducting hidden-treasure projects in existing buildings. For example, if a motor fails, the natural inclination in most organizations is to replace it with the same model. Failures are usually associated with urgent or perhaps emergency situations when the imperative is to minimize downtime. In that context, few organizations would reward personnel for taking time to research the availability of more efficient motors and to install an alternative model whose performance is unproved. On the other hand, few cash-strapped organizations would decide to replace a perfectly good motor simply because a more efficient model is available. Clearly, these opportunities are easiest to capture in new construction when it is possible, at least in theory, to do it right the first time. As subsequent chapters illustrate, sometimes this is easier said than done.

Not only are these projects complex from an organizational standpoint, they are often hard to fund. Finding an outside sponsor to upgrade infrastructure may be possible in some organizations. As we mentioned earlier, perhaps Wellesley's good fortune in having a benefactor who understood this point may inspire other generous and thoughtful donors.

Until outside funds materialize, colleges and universities will generally have to fund infrastructure system upgrades internally. As we said, it may be hard to argue successfully that functional equipment should be replaced because more efficient models are available. In some cases the efficiency gains are so significant that payback calculations will show that replacements pay for themselves in less than five years (due to reduced energy consumption). But if the organization is strapped for funds and has a choice of replacing failed equipment and that which is simply inefficient, most organizations will replace the failed equipment first.

One strategy for financing efficiency projects is to create a revolving-loan fund in which the initial savings from efficiency projects are used to finance more efficiency projects, and the postpayback savings accrue to the project sponsor (e.g., a department or school). With a loan fund of this type, replacement of functional but inefficient equipment can be justified. See a more detailed description of revolving loans later in this chapter.

Low-Hanging Fruit: The Easiest, Quickest, and Least Expensive Projects

Projects involving low-hanging fruit are those whose benefits can be realized relatively easily. An example of low-hanging fruit in most organizations is efficient lighting. Even though EPA's Green Lights program has been in place for over a decade, many colleges and universities with older buildings have lighting-efficiency opportunities that they have failed to capture. Replacing incandescent lightbulbs with compact fluorescent units, and replacing older fluorescent tubes with new, more efficient models, can be achieved at modest first cost with quantifiable long-term gains. Energy costs are reduced, and longer-lasting bulbs require less staff time for replacement. Insulation is another example of a project that usually falls into the low-hanging fruit category. Some exceptions exist.

For example, if an old building lacks insulation, it may also have wiring that is not up to code. In that case, the wiring must first be brought up to code before the space can be insulated.

Placing high priority on low-hanging fruit projects seems like a good idea for all climate action advocates. The problem is that for organizations such as Tufts that have been taking campus stewardship efforts for over a decade, few low-hanging fruit may remain. This is where a detailed knowledge of the organization's past environmental efforts is extremely valuable in helping inform the development of a climate action strategy.

Educational Projects

Educational projects are those with some practical utility, but whose primary short-term goal is educational. For example, our students have studied existing buildings on the Medford campus in an effort to learn which are best suited for small-scale solar installations, and groups of students have studied various aspects of converting a Medford central heating plant to a cogeneration facility. Several factors are likely to prevent the realization of either project. Our existing small-scale solar projects are great, but our next solar project will be incorporated in new construction (solar in curtain walls as well as rooftop arrays), and cogeneration is unlikely in the short term on the Medford campus due to competing priorities for capital, possible community concerns, and regulatory requirements.

As with other categories of projects in this discussion, it is important to understand the value of each project to the climate action mission. Having educational projects is entirely appropriate to an educational institution; however, not all of these projects will result in reduced emissions in the foreseeable future. If the time and other resources of climate action advocates are in short supply, it is worth spending time to periodically review the mix of projects for their emission reduction potential.

Other Attributes: Risk

New or new-to-the-place technology can fall into several categories, including large projects, educational projects, glamorous projects, or pilot projects. When technology is new or new to the place, the risk aversion of key decision makers is an important factor. Glamorous projects and pilot projects are a good match for new or new-to-the-place technology because expectations for a positive outcome can be managed. Large projects, especially those requiring large capital investments and large teams for decision making, may be the least fertile ground for experiments with new or new-to-the-place technology. However, it has to be noted that these are often the precise conditions for greatest reductions in emissions.

Off-the-shelf technologies are technologies that are proven either at your university or at a similar place. They are readily maintained, are proven, and deliver good results. In contrast, "top-of-the-shelf" technologies are new, pilot, or demonstration technologies where the given installation is the first. While top-of-the-shelf technologies are appealing because they often promise greater results, most universities are better off selecting technologies with a proven track record. In the long run, a high-performance technology that performs well is better than one with extraordinary theoretical performance but that does not perform reliably. Advocates should test (through references from similar installations) whether reliability is demonstrated.

Projects that are replicable, sustainable, and easy are the most sought after by climate change advocates, and it seems, the most difficult to identify. One project that falls in this category is the installation of VendingMisers®. At the outset, this project seemed like a classic "no brainer." Vending machines use a great deal of electricity, and when we suggested that fewer machines be located around campus, we were immediately greeted with a negative response from the campus organization that benefits from the proceeds. The VendingMiser® is a product designed to reduce the energy consumption of machines by activating select features with a motion sensor. At first, the no brainer turned into a near debacle when vending maintenance personnel disabled the devices installed by the Tufts Climate Initiative. Following extensive diplomatic efforts, the situation was resolved to the satisfaction of all parties. TCI staff wrote a piece for our website explaining how to avoid our mistakes, and it has been used by the manufacturer of the VendingMiser®, as well as people at other colleges and universities. So the system is working at Tufts, and it is working at other institutions. As we note in chapter 8, manufacturers have started producing vending machines that use a great deal less energy, but in the interim, there is a solution that is accessible and replicable.

Determining Priorities

With this complex array of risks and rewards it is not easy to set priorities. Some colleges and universities will select the path of least resistance, pursuing the low-hanging fruit and the educational projects along with the occasional glamorous project to keep the effort high profile and energized. Other organizations will have a climate action advocate who is willing to tackle large projects with great emission reduction potential, running the risk that the effort may be in vain if the project is not built due to factors outside the climate advocate's control. Some efforts will be driven primarily by the inventory, placing highest priority on the greatest sources of emissions.

Priorities may be easier to set after an initial project is undertaken. Our experience suggests that selection of an initial climate action project may reflect the advocate's view that a situation is ripe for action, or that a window of opportunity exists. For example, the initial project carried out at Oberlin College under the advocacy of David Orr was the conceptualization and construction of a high-performance environmental studies center. In his effort, Orr was successful in attracting donors (both financial and in kind) and engaging students and administrators in an ambitious effort to demonstrate the feasibility of the process as well as the design features. In our case, we sought modest success with quick results, the renovation of a small student residence to capture emission reduction opportunities described in the next section.

TCI's First Demonstration Project

Each year the Tufts Facilities Department selects buildings for maintenance-related capital improvements as part of Tufts' deferred maintenance plan. The Facilities Department in 1999 suggested that the maintenance activity could be an opportunity to introduce energy- and environment-related improvements. The Tufts Climate Initiative asked a graduate student group from the Department of Urban and Environmental Policy and Planning (UEP) to develop a set of recommendations that could be translated into action. This project was part of a required core course in which UEP student teams solve problems for an array of clients.

Working with the Tufts Energy Manager and TCI, the UEP students chose the university's French house—also known as Schmalz House—as the focus of their study based on the availability of energy-use data for the building, its inclusion in the summer maintenance plans, its potential for improvement, and its lack of occupancy during the spring 1999 semester. Schmalz House is a three-story residential wood frame building with living space for twelve students who form a French-speaking community. Prior to the maintenance work, Schmalz House had no insulation, inefficient lighting, two oil-fired boilers, and two inefficient gas-fired hot water tanks.

The study performed by the graduate students suggested that the university should insulate the walls and roof, install a high-efficiency gas-fired boiler, improve heating controls, replace the top-loading washing machine with a front-loading unit, replace the existing refrigerator with an Energy Star–labeled model, and explore the feasibility of a solar hot water system.

TCI adopted the students' study as the foundation for a demonstration project and worked with Tufts Facilities, who implemented a number of the suggestions put forth in the students' report, as well as some additional items. TCI provided technical assistance and funded a portion of the climate change reduction–related improvements at Schmalz House.

The project was implemented in two phases during the summers of 1999 and 2000. Phase 1 (summer 1999) included the introduction of

- A solar hot water system with Btu meter
- High-efficiency lights
- Auto-dimming ballasts and other advanced lighting controls
- An Energy Star refrigerator
- A front-loading washing machine

Because the UEP students worked on the project in the first year of their two-year master's program, they could see the changes enumerated above when they returned to campus in the fall. Phase 2 (summer 2000) activities included

• Replacing the dual oil-fired boiler system with a high-efficiency gas-fired boiler
• Converting to all-hot-water heat distribution with baseboards
• Insulating walls and roof
• Connecting the existing hot water tanks and solar tank as a zone off of the boiler

The changes were expected to reduce energy use and emissions:

• The Energy Star refrigerator uses 20 percent less energy than a typical refrigerator.
• The front-loading washing machine uses nearly 50 percent less water and 35 percent less energy per load than a top-loading unit.
• The high-efficiency lighting will use nearly 25 percent less energy than the previous lighting system.
• The solar hot water system is expected to offset approximately 20 percent of the building's water-heating needs depending on the amount of water-use reduction resulting from the new washing machine.
• The more efficient boiler will reduce total heating energy and the switch from oil to natural gas will reduce emissions of climate-altering gas per Btu.

Consistent with TCI's strategy for using the campus as a learning laboratory, this same house was the focus of a student project in another course in spring 2005. See chapter 9 for an evaluation of these emission reduction actions.

Funding Climate Action

Goals, money, and priorities are linked in a complex network. Some projects are easier to fund than others. Whether a climate action advocate should place priority on projects that are relatively easy to fund is a strategic decision, and one that advocates may wish to revisit on a periodic basis. In this section we explore factors that may enhance the success of climate action advocates, all of which relate directly or indirectly to funding. Not all of these factors are equally salient for all situations, so knowledge of your institution will help you decide on an approach that will work.

Long-Term vs. Short-Term Thinking and Investing

One of the attributes of climate change that makes it both appealing and daunting is that we will have to reframe the way we think about decisions if we are to act in favor of emission reduction. Life-cycle thinking is essential. The Society of Environmental Toxicology and Chemistry (SETAC) continues to develop the science of life-cycle assessment in which the full range of effects to the environment are considered from raw material extraction to ultimate disposal.[3] The life-cycle approach is not unique to climate change, and in fact is used in a range of industry applications. Increasingly, manufacturers think about end-of-product-life issues at the design phase to facilitate disassembly and resource recovery. Life-cycle costing issues are also entering the decision making of an increasing number of consumers, particularly as gasoline costs rise and consumers calculate the long-term savings of operating a fuel-efficient car.

Especially in the area of energy efficiency, the first costs of a very efficient item are often higher than for the less efficient item. It is during the operational phase that savings are realized. This is true of a range of consumer products, including compact fluorescent lightbulbs and energy-efficient front-loading washing machines. First costs may also be higher for energy-efficient motors and boilers, making first costs for high-performance buildings higher than for their less efficient counterparts. However, the costs of operating the green building are lower (because less energy is used), and the total costs over the building's life cycle are lower than for conventional designs.

The assumption is that academic institutions will be in existence for the long term, and thus should embrace life-cycle costing with great enthusiasm. In practice, however, things may be different. If a new building is being planned and the first costs of an extremely efficient heating system are 40 percent higher than the first costs of a conventional heating system, the university may install the conventional system even though the efficient system might pay for itself in less than five years. The reason for this seemingly irrational decision is that the additional 40 percent for the more efficient system may simply not be available in the budget for the new building. We believe it is entirely feasible and highly desirable for academic institutions to create loan funds or other mechanisms to cover the incremental first costs associated with energy efficiency; however, key decision makers may not be aware of the benefits to the institution of taking such an approach. This is where a climate action advocate's intervention may help reframe decisions.

Long-term thinking considers the effects of climate change on the physical infrastructure of the institution as well as its operations (see chapter 9). Long-term thinking will consider energy capacity, energy sources, reliability of energy systems, and environmental permits. A university that fully embraces climate change action will expand this thinking to less direct and quantifiable impacts and will include the improved image for prospective students, the leadership value among donors, and the visibility for faculty and alumni. At Tufts, we are now working to develop this thinking.

One example of long-term thinking relates to holdings in the university's investment portfolio. A 2002 report by CERES called *Value at Risk: Climate Change and the Future of Governance* examines the implications of climate change for company directors, fund managers, and trustees and concludes that they will be shirking their fiduciary responsibility if they fail to take climate change into account in evaluating investments.[4] We view this as a compelling argument for colleges and universities to scrutinize their endowment portfolios and consider divestment from companies whose core businesses are large-scale generators of greenhouse gases, such as utilities relying on coal. An alternative to divestment is shareholder activism. Active shareholders vote selectively on resolutions and engage in dialogues with company management on issues of concern. Historically, institutional investors such as pension funds and college and university endowments have not voted their shares, and thus have ceded their power to companies' management. A college or university concerned about climate change and financial risk will vote in favor of climate-related resolutions and will communicate with management in companies vulnerable to climate change to determine whether they can shift to a more sustainable business model. In either case, the academic institution will benefit from considering the long-term implications of its investments and from acting to protect investment value.

Funding Projects

Finding money to undertake climate change actions may be relatively easy at colleges and universities where decision makers readily embrace life-cycle thinking, where climate action goals dovetail well with strategic priorities of the institution, and where capital expenditures for infrastructure improvement are reasonably available. We have been challenged to support our climate action efforts, and offer suggestions for those in situations comparable to ours.

The Cash Catalyst Model This is an approach that uses grants or other external capital to fund the difference between the first cost of energy-efficient systems and conventional systems. The Tufts Climate Initiative has been extremely fortunate to have the support of the Kendall Foundation and the Rockefeller Brothers Fund (RBF). Aside from modest support for part-time staff and students, we use funds from the Kendall Foundation and RBF to leverage expenditures by the university. We think of this as a "cash catalyst model." This approach was used when the university planned and built a new Wildlife Clinic at the School of Veterinary Medicine. TCI funded the difference in first cost between conventional and very efficient air handling equipment for the building. Energy use and payback are being monitored. The assumption is that with the positive experience, the university will eventually use more efficient equipment as a matter of routine.

Energy Service Companies (ESCOs) An ESCO, or Energy Service Company, is a business that develops, installs, and finances projects designed to improve the energy efficiency and maintenance costs for facilities over a seven- to ten-year time period. ESCOs generally act as project developers for a wide range of tasks and assume the technical and performance risk associated with the project. Typically, they offer the following services:

- Developing, designing, and financing energy-efficient projects
- Installing and maintaining the energy-efficient equipment involved
- Measuring, monitoring, and verifying the project's energy savings
- Assuming the risk that the project will save the amount of energy guaranteed

These services are bundled into the project's cost and are repaid through the dollar savings generated.[5]

Colleges and universities use ESCOs for heating, ventilating, and air-conditioning engineering services and compensate companies for their work with an agreed portion of the early energy savings from renovations and equipment upgrades. Tufts has used this approach in large buildings in our health sciences campus in downtown Boston. The Commonwealth of Massachusetts has used the ESCO approach extensively to fund energy-upgrade projects; because these projects fund themselves via energy savings, legislative appropriations are not required. ESCOs are experienced in energy efficiency and have an incentive to generate savings for the building owner. However, because ESCOs are motivated by profit, they may overlook projects that have a longer-term payback. Relying on ESCOs alone can have another downside: the university may forgo opportunities to undertake maintenance at the same time. If a college or university has engineering staff trained for this type of work, the cost to the institution may be less and the systemwide benefits greater if the work is done in-house.

Revolving-Loan Funds Such funds are an attractive strategy for encouraging energy-efficiency measures. An initial fund is established and principal repayments of loans are put into the fund and made available to other borrowers. In a climate action fund, savings generated by energy efficiency are repaid into the fund and then used to finance new energy-efficiency projects.

The Tufts loan fund was created with savings from energy rebates that were offered by utilities and accrued over the years; however, other initial funding mechanisms such as grants or one-time budget allocations are possible. Another strategy is to create the loan fund from a portion of the endowment. We think that investing part of the endowment in campus energy efficiency is a brilliant idea. At present, energy efficiency is yielding a higher rate of return than many conventional investments. Regardless of the mechanism used to create the fund, the operation is simple. The project advocate calculates the savings that will accrue from using the new, more efficient equipment, and then repays the purchase price of the equipment from the operating budget over an agreed period. When loans are repaid, funds are made available for subsequent projects.

Rules for eligible projects may vary depending on the risk profile of the college or university and the people making decisions about loan-fund management. Most projects at Tufts are expected to have a five-year payback (or less). Five years may sound like a long time in the context of some corporate decisions; however, it is a relatively short time in the life of an academic institution, or even a homeowner.

Full Cost Accounting This strategy is designed to improve decision making. In general, full cost accounting reveals direct and indirect costs, contingent liability costs, and other less tangible costs such as social and environmental costs associated with a particular course of action. If a college or university were to introduce full cost accounting for energy, it would increase the unit price to each department, laboratory, or other user. Full cost accounting reflects the fact that it costs more to deliver a unit of energy to a campus building than the cost of the fuel used. Additional costs experienced by the college or university include capital costs of systems, distribution and system maintenance costs, as well as administrative costs such as conducting negotiations with utilities and managing contracts with service providers. Typically, all of the additional costs will be covered in an undifferentiated overhead charge. With full cost accounting for energy, the institution takes the additional costs out of overhead and allocates all the energy-related costs to energy users on a per unit basis. This revised institutional pricing increases the price of energy to the user and is intended to provide an incentive to reduce consumption.

Tufts does not have full cost accounting for energy. In fact, as we mentioned earlier, Tufts does not make individual departments responsible for energy costs. Energy is addressed at the school or campus level. By making energy use an overhead item, researchers in political science and economics are subsidizing the research of their colleagues in physics and electrical engineering, whose energy demands are substantially greater. This may or may not be reasonable for the institution. There are a range of practical barriers to allocating energy costs to departments. For example, some departments share buildings, so an allocation model would have to be developed. Without sophisticated metering, actual consumption could not be used. If people in two departments collaborate on a research project, whose energy costs should be used in the budget? On the other hand, if all of the energy bills are paid centrally, what is the incentive for groups or individuals on campus to begin turning off their computers and other equipment when not in use, and ending other wasteful practices?

Availability of Funds The availability of financing for priority projects may be constrained. As we indicated earlier, this may be particularly true if priority projects consist primarily of the hidden treasures such as steam traps and major capital projects such as upgrades to the central heating plant. These projects are unlikely to attract sponsors such as foundations and individual donors, in part because they are so mundane that there is a sense that the college or university should be undertaking these efforts as a routine matter.

A pitfall for climate action advocates is that external grants and foundation funds are much more readily available for education projects and other activities that raise general awareness of climate change but do not in themselves reduce emissions of heat-trapping gases on campus. As we have said, we undertake education projects, but in doing so we are mindful that there are trade-offs because the time we can devote to climate action is limited.

The Challenges

Societal Challenges

One of the greatest challenges in advocating for climate action derives from our place in a culture of affluence and our patterns of casual consumption. This is not a factor that we explore in great detail at the Tufts Climate Initiative; however, it is one of several aspects of climate change that deserve attention in the academic community. Instead of examining affluence, we generally focus on managing its consequences with climate change in mind. We recognize that this approach has drawbacks, but also recognize that there are limits to our ability to add meaningfully to a discourse that is already fraught with complexity, particularly when our focus in on reducing emissions.

Governmental Challenges

Another challenge is associated with the approach being taken by national governments to address (or fail to address) climate action. We take the view that there are several "no regrets" measures that if implemented by national governments could have multiple positive effects on economies, on local environments, and on global emissions of greenhouse gases. These measures include extremely ambitious fuel-efficiency standards for vehicles (including those for air transport), efficiency standards for motors and for a wide range of appliances, and lighting and air-conditioning standards for newly constructed public and private buildings. Absent measures of this type, the efforts made by colleges and universities to save money and reduce emissions will be a great deal more difficult than they should be.

Institutional Challenges

Given our limited resources, we cannot take advantage of all the climate action opportunities on campus. What makes the most sense? If we have only half an hour, do we spend it with a reporter from the student newspaper who wants to write a feature on climate action on campus, or do we research sources of inexpensive compact fluorescent lightbulbs and replenish our dwindling supply? The answer, of course, is that we spend time with the student reporter, hoping that in the long run the article will inspire many others to action.

Perhaps the largest mistake we have made (and probably continue to make) is to take actions that are relatively easy and to postpone the really hard actions for the future. For example, it is relatively easy to sponsor or cosponsor an event that provides technical information on high-performance buildings or that features low-emission transportation options. It is much more challenging to reach agreement with the university on a plan to switch to green power.

In many cases, an action is challenging because it requires taking an approach that is different from the norm. The capacity to change, to take risks, and to be open to innovation will need to be developed as we move toward the university of the future.

Taking climate action at a college or university also can be challenging because each member of the community makes decisions that affect emissions of greenhouse gases. By examining links between receptive decision makers and information on emission sources from the inventory, an effective climate action strategy can be developed. The strategy can be opportunistic and focus primarily on the glamorous projects to attract attention, or it can be planful and oriented toward influencing large projects with significant implications for campus emissions. Because there is such wide variety of colleges and universities, there will be vastly different responses to efforts by climate change advocates. Some advocates will find their ideas are welcomed by a receptive and responsive organization. Other advocates will find that they need to work hard to cultivate a positive climate for emission reduction.

Developing a Climate Action Mindset

The climate advocate at some colleges and universities might initially feel like a lobbyist who has no money to distribute, no expense account for lavish meals, and no favors to grant. But in reality the climate action advocate can offer a great deal. We outline some approaches here.

Be Useful

A good strategy is to become useful in discussions that can influence climate-altering emissions. For example, you can offer to take minutes in a meeting, or can offer to research products and provide information to the decision-making team (whether or not the climate advocate is a formal member of the team). If the discussion reveals uncertainty about technology that is new or new to the place, strategic research and information gathering may be very helpful. Learning that the technology is used successfully by competing institutions may be just the sort of comfort factor that decision makers seek.

Use Questions

Ask questions to inform climate action advocacy. In most academic settings, a carefully posed question will not be automatically taken as hostile; however, it remains important for the climate action advocate to read all the signals carefully. Posing a probing question can be an effective advocacy technique, though it is one that can suffer from overuse. Questions can reveal information about the process, provide an opportunity for validation, and yield information advocates can use to follow up. Ask specific questions such as:

- What is the efficiency?
- Where else is this technology used?
- What assumptions are being made?
- May I please see the calculations?

Your questions may generate new thinking about a project, or lead to considerations that had been overlooked. At times your questions may yield information useful for future projects, as our questions did at Tufts when we discovered our assumptions were wrong as to where the authority for many climate-sensitive decisions lay. We have also learned that some decisions critical to the campus emission profile are not thought of as important decisions. Alternatively, some critical decisions were considered years ago in another context and options that would have reduced emissions were rejected for reasons unrelated to the environment or to climate change. Learning the history of these decisions by asking questions of many of the people involved helps us decide whether and how to ask the institution to revisit decisions.

Evaluate Answers

Evaluate answers, particularly when they are unexpected, as an important element of strategy formulation. There will be times when answers to questions reveal procedures or interests of which you were unaware. For example, in the very earliest stages of discussing a new building, it is always useful to learn whether an architect has been selected. Because some architectural firms are actively committed to green or high-performance buildings, firm selection can be an important opportunity for the climate action advocate. We have learned that in some cases, the major donor for a building may express a strong desire that a particular architect be used, or may favor design features that have undesirable implications for energy use. In these cases, you need to think carefully about the resource and political implications of trying to convince decision makers that the donor's express wishes be questioned.

An example of an emission-intensive architectural feature at Tufts is a heated outdoor walkway linking the top floor of a hillside building to the upper campus. Heating the walkway was not the only way to ensure that it is free from ice, snow, and rain; indeed, a simple rooflike covering would protect pedestrians from inclement weather.

In general:

- Make sure your question is answered.
- Do not accept vague answers, but do accept the possibility that the person giving the answer may simply not know enough to be specific. Ask other people.
- Be wary of these answers:
  - "We've always done it that way."
  - "We tried it in the 1970s or 1980s and it didn't work."

Follow Through

Follow through on commitments to establish your credibility and to advance climate considerations in decision-making processes. While this may not be the case on all campuses, we discovered at Tufts that the parts of the organization most directly able to realize emission reductions are also the parts of the organization that are most strapped for resources. This means, for example, that if we agree to schedule a meeting and send out an e-mail in advance to remind participants, we are making time for a person in the facilities department to take an action, however small, that will reduce emissions. In just a few spare minutes a facilities person can schedule someone with a tall ladder to install compact fluorescent bulbs in a chandelier or arrange for an energy audit of the president's house.

Be Respectful

Be respectful, even when it is difficult, and you are more likely to be respected. It may be challenging to be respectful when you observe work that is performed poorly by others or decisions that appear irrational or ill-informed. When you work from the inside to transform an organization's decision making you will have to assess when being respectful of the status quo is a useful strategy, and when it is time to insist on change. Being respectful of individuals is simply good manners and helps build relationships. This is particularly the case at a college or university where employee turnover is low. If you are a member of the staff or faculty and you are successful in changing the way decisions are made, there is every reason to assume that you will be dealing with the same people both before and after the transformation in decision making. Remember that the transition to the university of the future is an enormous challenge that requires a long-term commitment.

Be Realistic

Be realistic about the pace of climate action on campus. You cannot expect to transform an organization overnight; however, regular progress is a reasonable expectation. The definition of what is realistic varies, and as an advocate you will shape expectations. Some organizations are thrilled to achieve a vague general awareness of climate change and to inspire emission reductions from efficient lighting (harvesting low-hanging fruit). Other colleges and universities have clear goals such as a Kyoto commitment or carbon neutrality and seek to quantify and track progress toward that goal.

One strategy that you can pursue is a periodic reality check or evaluation. Reality checks can be formal or informal and internal or external. An internal and informal reality check might involve asking key faculty, students, and operations staff how they think the climate action effort is going, what can be done more effectively, and what is not being done that should be. An informal external reality check can consist of asking a climate action advocate at another college or university in the area or with similar goals about particular challenges and strategies for addressing them. Depending on your professional network, similar questions might be posed to climate advocates in state or local governments or at the Environmental Protection Agency's regional office.

More formal reality checks may depend on your accountability structure for climate advocacy. For example, if you have a steering committee composed of faculty, students, and staff, then periodically discussing the pace of action and learning about committee expectations can be extremely valuable. If your climate action program receives external financial support, a periodic evaluation of the rate of progress toward reaching goals can help refine not only what is realistic, but also what the granting organization views as priority actions. We talk more about the importance of evaluation in chapter 9.

Expect Success

Expect success in advocacy efforts and plan your next move. Because of the increased media attention climate change is receiving, it is possible that you will have a variety of people—including students, faculty, alumni, and staff—asking what they can do to help. It takes time to manage volunteers; however, with advance planning, you can generate a list of small-scale projects to pursue when new volunteer resources become available. Also, it is useful to have on hand a list of ambitious project descriptions for prospective supporters when your efforts rely on external grants. You never know when someone might call and ask how you would spend a few million dollars to reduce emissions.

Be Concrete

As part of its campus stewardship program in 1990, Tufts was among the first universities to develop an environmental policy. Although the policy has been used to reinforce actions—for example, helping to justify energy conservation measures—it has not had an obvious impact on day-to-day decision making. The goal of meeting or beating emission reductions associated with the Kyoto Protocol appears to be having a different impact on decision making. Because the goal is quantifiable, and because it is possible to assess progress toward the goal on a regular basis, the commitment already is being used as a rationale for influencing decisions such as efficient-appliance purchase.

Depending on the context, it may be important for you to offer very specific suggestions. For example, in suggesting that emissions be reduced during building renovation, it may be useful to identify opportunities such as an efficient boiler and lighting upgrades and offer examples (including makes and model numbers) of products that are efficient.

Know Your Limits

Many of us realize that we have reached the limit of our knowledge when the conversation turns to boiler performance, steam traps, variable-volume air handlers, and insulation choices. Fortunately, there are experts who will help sort out all of these details and many more. Despite our having walked through and studied many buildings and having developed some skills in this area, we are not experts. We ask experts to walk through buildings with us to establish a sense of what magnitude of emission reduction can be achieved with a renovation, and we rely on experts to analyze plans as well as to advise on equipment selection and installation and a host of other activities. In designing a new building, a high-performance building expert and a solar expert will work as part of an integrated design team.

Once people have been sensitized to concerns of energy use in buildings, they may become aware of rooms that are uncomfortable and may draw erroneous conclusions. For example, in the fall and spring New England experiences swings in temperature from day to day. A few days in which high temperatures are 75°F may be followed by a day in which the high may be 50, followed by a return to warm days. These are the conditions that make life difficult for facilities personnel. On the day when the high is only 50, it is common to hear complaints that energy is being wasted because the air-conditioning is on in meeting rooms and classrooms, when in fact it is not. The problem is that fresh air being vented into the room is simply not being heated. And although the experience can be very uncomfortable, it is not due to wasteful use of air-cooling equipment. Instead, these uncomfortable conditions may be an indicator that energy is being saved rather than wasted.

Summary

Actions that reduce emissions on campus can be classified in several ways. Most emission-reducing actions involve: (1) increased efficiency, (2) increased use of alternative fuels or green power, (3) fuel switching, or (4) reduced demand through changed behavior and expectations. In developing an action plan for your campus, you may want to consider a strategic mix of projects to include some that are low-hanging fruit—easy, quick, and inexpensive, some that are glamorous but might have a modest effect on emissions, and some that are hidden treasures—low-visibility projects with high potential for reducing heat-trapping gas emissions. In the next two chapters we focus on emission reduction in buildings and provide examples of many of these measures.

6

Buildings and Climate Change Action

College and university climate-altering gas emissions result largely from the burning of fossil fuels for heat, hot water, and the generation of electricity. Therefore action for climate change is most effective if addressed in campus facilities including buildings, grounds, central heating and power facilities, water heating, and other building functions. Climate change action will encompass all aspects of university facilities: campus planning, including facility and infrastructure planning (see chapter 9); new-building design and construction; existing facilities; maintenance; and central facilities. Effective climate change action will address these emissions directly using the types of climate change action described in chapter 5—efficiency, fuel switching, green power, and alternative fuels.

This chapter focuses on ways to make the buildings you manage, design, live in, and work in less burdensome to our planet. We recommend that you become familiar with the wide variety of building resources (see appendix B), recognizing that there is a vast technical literature as well. A more detailed description of specific building-related opportunities follows in chapter 7.

Although we have experienced some real successes in our efforts at Tufts, we have also experienced some frustration. Many of the disappointments were earned through attempting to integrate apparently incompatible ideas of building systems with university decision-making systems, timelines, and budgets that had the potential for compromising positive outcomes for emission reductions. Our discussions will extend, however, far beyond what has been done at Tufts, since we are really just beginning.

Why Buildings and Facilities?

The energy services required by residential, commercial, and industrial buildings produce approximately 43 percent of U.S. carbon dioxide (CO_2) emissions.[1] As a consequence, modifying existing buildings and building systems and maximizing reductions in new buildings are key elements in any comprehensive emission reduction strategy. This is particularly true at colleges and universities, where typically most of a college or university's greenhouse gas emissions come from heating buildings, providing hot water, and generating electricity needed for cooling buildings as well as powering lights, fans, motors, and other building systems. We are born in buildings and we live, learn, eat, sleep, and often work and play in buildings, so we all have a great deal of experience with buildings.

EPA estimates that nationwide, Americans spend 90 percent of their time indoors.[2] Yet most of us have little idea how buildings are built or how they operate. Most of us take buildings for granted—until they are too hot, too cold, or leaking. We rarely think about a building unless it has striking architecture or some other unusual feature. Building occupants rarely stop to think about how systems work to make the building functional and comfortable or what the impact of those systems on people and our environment might be. Buildings provide opportunities to teach: How do the lights get power? How does the building get heat? What fuel is used for heat, hot water, or power? Where does the fuel heating the place come from? Where does the waste go? How is the building cooled? What are the origins of the wood, steel, paint, and carpet? Unfortunately, few engineering schools educate the next generation of mechanical, structural, renewable-energy, or systems engineers to develop the breakthroughs needed.

Buildings are visual symbols of a school's legacy—its ideals, strength, wisdom, art, and athletic and intellectual achievement. College buildings, some spanning a century or more and bearing the names of famous alumni, past presidents, or generous donors, inspire each new generation to maintain and extend the institution's legacy and values. It is time that these legacies are more than just bricks and mortar.

There is a very strong fiscal and environmental rationale for focusing climate change action on buildings and other facilities in any organization, but particularly in academia. At universities and colleges, buildings are typically owned and operated by the institutions for many decades, even a century or more. Tufts recently demolished a "temporary" psychology building that had been in place for 107 years! Most institutions plan to be in operation 50 years or more, so long-term investment in existing buildings is a sound financial decision, and careful attention to energy efficiency in new buildings will reduce operational costs over the building's life.

Many climate change measures can have other benefits, including improved light quality, improved ventilation, and reduced noise. These benefits can improve productivity and learning. However, this link is rarely made, even in buildings where the obvious academic material is relevant to the building systems, such as engineering or environmental studies. Less obvious academic connections are nearly always overlooked, such as in departments of education (natural light can raise test scores)[3] or management classes (healthy buildings can reduce absenteeism and increase productivity).[4]

University buildings serve a wide range of functions—dining halls, residences, offices, classrooms, laboratories, and storage. Many buildings house multiple functions and most campus building uses evolve over the years. One of Tufts' oldest buildings, Ballou Hall, is still going strong. When Ballou Hall opened in 1852 it housed the chapel, student rooms, and faculty offices. Now 150 years later, it has offices for the president, deans, and the graduate school. The advent of new technology, research developments, and changing university calendars (e.g., more summer programs) often result in multiple renovations or building uses that are far from the original design intent. Buildings are complex—boilers, air conditioners, air handling, elevators, lighting, controls, wastewater, and fire protection are just a few of the systems that must work together to provide comfort and safety for building occupants. A vast array of professionals specializing in various aspects of building design, renovation, and maintenance are responsible for carrying out this complex design and operation.

Applying Climate Change Action in University Facilities

The overall climate action strategy for a college or university will be informed by the emissions inventory, an understanding of the organization's existing building stock, and the institution's building plans. Climate change action affects all aspects of university facilities: campus planning, including facility and infrastructure planning; new-building design and construction; renovating and maintaining existing facilities; and management and maintenance of central facilities. Effective climate change action will address these emissions directly using the four major types of climate change action—efficiency, fuel switching, green power and alternative fuels, and from taking an integrated approach. Table 6.1 shows examples of each action and indicates how they apply to the stages of facilities management. It is important for climate change advocates to realize that the list of possible projects is nearly endless, that most of the most effective projects are highly technical, and that often expenditures (large or small) are needed to accomplish these measures.

Efficiency

Climate change action in facilities must include energy efficiency—using less energy to deliver the same level of service. This includes using less electricity in building systems: using less electricity in systems such as chillers, motors, fans, laboratory hoods, and lighting; improving the efficiency of building heating and cooling systems and reducing the energy needed to heat hot water; and tightening the building envelope (walls, windows, doors, roof) as well as reducing penetrations of the building shell or thermal envelope. (See box 6.1 for information on electricity units.) Regardless of the other strategies pursued, energy efficiency should be a part of any overall plan. Efficiency should be mandatory, especially at important decision and investment points such as in new construction, in renovations, and when a programmatic change occurs (for instance, when a new laboratory is needed or a department outgrows its space and is moved to another building). Energy-efficiency opportunities exist throughout all aspects of facilities management and should be a routine part of the management process. Efficient systems can be smaller and cheaper, use less fuel, and take up less space. Some buildings offer huge and immediate opportunities to reduce energy use from lighting or improve the efficiency of the boilers or hot water heating. Other buildings already operate quite efficiently, and more intensive study is required to find opportunities to improve efficiency.

Energy efficiency can include utilizing a wide range of technologies that use less energy. However, efficiency can also include new or

Table 6.1
Examples of climate change action in facilities

	Planning	New buildings	Renovation	Maintenance	Central systems
Efficiency	Locate buildings near central steam plants. Building orientation can reduce heat gain or capture it.	Improved insulation can reduce heating and cooling needs. Glazing can maximize daylight and minimize solar gain. High-efficiency heating systems save energy.	Lighting upgrades can improve efficiency and reduce cooling loads. Replacing energy using equipment with more efficient models can reduce energy use.	Strong maintenance programs can identify building inefficiencies on a routine basis. Cleaning steam traps, air filters, light fixtures, etc. improves their efficiency and/or effectiveness.	Combined heat and power systems can generate electricity and steam concurrently.
Fuel switching	Fuel cells may provide emergency power or high-quality reliable power	The choice of fuels has implications for emissions. Natural gas has lower emissions/ Btu than oil, for example.	Switching from oil to natural gas will reduce emissions.	Cleaner fuels may reduce air-permitting obligations. Cleaner fuels may be easier to manage and handle.	Switching from oil to natural gas in boilers can reduce emissions. Dual-fuel capacity provides an opportunity to switch based on price or other factors.
Green power and alternative fuels	Building-integrated systems can be designed into new construction.	Solar thermal (using sun to heat hot water) or geothermal can reduce use of fossil fuels.	Solar thermal and PV can be retrofitted on existing buildings.		Green power, such as electricity generated from wind or hydro, can reduce emissions.

Box 6.1
What's a watt?

Watt
Watts are a basic measure of electrical power. The watt is named for James Watt (1736–1819), a Scottish inventor.

A *watt-hour* is one *watt* of power for one hour. A 100-watt lightbulb burning for one hour will use 100 watt-hours of electricity. One watt-hour equals 3,600 joules of energy.

Kilowatt-Hour
A *kilowatt-hour* (kWh) is the standard unit of measure for electric energy. One kilowatt-hour is equal to 1,000 watt-hours. The kWh is the most common measure of electricity use.

Megawatt
A *megawatt* is 1,000 kilowatts or 1 million watts; it is a standard measure of electric power plant generating capacity.

Megawatt-Hour
A *megawatt-hour* is 1,000 kilowatt-hours or 1 million watt-hours of electrical energy or 3.6 billion joules.

additional systems, such as heat recovery, improved controls, or reduced hours of run time. Addressing energy efficiency in campus buildings is nothing short of a monumental challenge, particularly when there are many old buildings. Often the most creative measures must be funded from outside sources or funded from available capital reserves and paid back from savings. But these sources of funds are often the first to be tapped when operating and capital budgets are tight, since funding maintenance and efficiency is not glamorous, new, or publicity-worthy in its traditional form.

While many colleges and universities have a dedicated energy manager who oversees the procurement of fuels, electricity, and energy services, attention to improving energy efficiency should be the focus of all members of a facilities team. Climate change advocates can help to identify efficiency measures, to determine ways of funding them, and to look for opportunities to improve and maximize project effectiveness. However, energy-efficiency projects usually require specialized engineering and knowledge of complex building systems to be successful. A more comprehensive list of specific opportunity areas for buildings follows in chapter 7.

Fuel Switching

Fuel switching refers to a change in energy sources—that is, changing from electric to gas heat or from oil to natural gas. This change can have significant positive benefits. By changing to a fuel with lower carbon dioxide emissions, your institution's overall emissions will decrease. Often, when a fuel switch occurs, an older piece of equipment (water heater, boiler, and so on) is replaced with a newer, more efficient model that can lead to even greater emissions reductions and fewer maintenance problems.

Fuel switching can also be an opportunity to save money. For example, a dual-fuel boiler (such as one that can run on oil or natural gas) can be switched to natural gas when price of gas is low relative to oil. This equipment also allows the option of switching back if natural gas becomes more expensive than oil. This latter scenario is a problem for climate advocates because burning oil releases more heat-trapping emissions per Btu than natural gas.

Bowdoin College recently switched its central campus steam plant from #6 fuel oil to #2 fuel oil. By converting from #6 fuel oil to #2 oil, Bowdoin produces 57 percent (forty-six tons) fewer emissions of sulfur dioxide and particulate matter a year and has decreased nitrogen oxide emissions by 77 percent annually (sulfur dioxide and nitrogen oxide are the primary causes of acid deposition). Reduced emissions also save the college approximately $570 each year in emission fees to the State of Maine. Additionally, the switch to #2 oil reduces operating costs due to decreased maintenance requirements, fewer fuel additives, and greater combustion efficiency.[5] Bowdoin's fuel switch has also reduced greenhouse gas emissions at a calculated rate of slightly less than 4 percent.[6]

Green Power and Alternative Fuels

Green power is electricity generated from fuels that do not deplete the earth's resources. Solar and wind energy are the most commonly thought of green power or renewable-energy sources, but tidal energy and hydropower are other examples of electricity-generating sources. Universities may opt to purchase green power (see chapter 8) or to generate power on campus. On-site generation may be central (see the discussion later in this chapter) or distributed (located where or near where it will be used).

Alternative fuels are generally considered to be alternatives to petroleum or coal and include renewable-energy sources. These can also be used for heating hot water, creating steam for heat, or absorption cooling. Examples include biomass, wood pellets, and geothermal energy. Numerous technologies can generate electricity or heat or both at the same time from conventional and alternative fuels.

Solar Thermal Solar thermal systems utilize the sun's energy to heat hot water. The process uses the sun's radiant energy directed on a solar thermal (often called solar hot water) panel. A fluid (usually water or glycol) inside the panel is heated and then circulated to provide heat to domestic hot water. Solar thermal systems make sense for residence halls or other buildings, such as gymnasiums, with high hot water usage. At Tufts, a small-scale installation showed that we reduced hot water heating costs by 50 percent even in a residence that is not used in the summer. As a result of the successful demonstration, a new Tufts residence hall will have a large solar thermal installation. In that application savings will be significant since the building will be occupied, at least partially, during summer break.

Some states, such as Maine, provide incentives for the installation of solar thermal systems. These incentives are generally in the form of favorable utility rates, grants from a state agency, or tax credits.

Solar thermal systems are simple and require little maintenance. However, like any system they do require some care. Many campuses installed solar thermal systems in the 1970s that sit idle now, often for lack of knowledge about them or for lack of replacement parts. Old, nonworking systems may be a barrier to future use that will need to be addressed on some campuses. Maintaining new systems is critical to their future success.

Photovoltaics Photovoltaic (PV) panels absorb the energy of the sun and convert the energy into electricity. PV panels and their associated equipment can be installed on some rooftops, as stand-alone systems, and as building-integrated (built right into a new building) systems. New technologies from rooftop units to PV shingles are emerging rapidly. In most university applications, the electricity will be used by the building and excess electricity will feed into the grid. These "grid-connected" systems do not require batteries to store electrical power, but simply use the building's or university's demand to use the power as it is produced. In most cases, building-mounted or building-integrated systems will be insufficient to meet all the needs of a university building even when the building is unoccupied because of the large space-to-roof ratio and because most buildings remain quite energy intensive even when they are not fully occupied.

PV systems generate electricity without any heat-trapping gas emissions. They also generate power most effectively at times when power demand is usually the greatest—hot, sunny, summer days. However, PV systems do not produce power all the time, so it is unrealistic to expect to power a building entirely by sunlight unless battery storage is provided or the building operates very unconventionally. PV systems are also visible and provide an educational opportunity to discuss electricity, environment, power use, and generation. Once installed, PV systems are relatively maintenance free. Monitoring should accompany the systems to troubleshoot problems as they arise.

Unfortunately, the current generation of PV panels is still expensive to make and converts relatively low amounts of sunlight into electricity, which makes them costly compared to other forms of electricity generation. Several states, including Massachusetts and Rhode Island, have aggressive programs to help subsidize PV and other renewable installations. The next-generation PV panels are likely to be less expensive and/or more efficient. Innovations from combined PV and solar thermal systems will also lead to greater efficiency. Nonetheless, the increasing price of electricity and the increasing value of peak power (power used or needed when demand is highest) are improving the cost-effectiveness of solar systems. The emerging market for the "green attributes" of solar (see chapter 8) increasingly makes electricity generation from solar a more cost-effective alternative using conventional cost-benefit calculations. One challenge in evaluating these costs and benefits is to accurately predict the future cost of power.

Wind Wind energy is generally the most cost-effective approach to generating green electricity currently available. Wind farms are going up around the country, with California and Texas leading the way in generation capacity. Modern wind turbines can stand as high as 300 to 400 feet with blades that are nearly half the length of a football field. These industrial-scale turbines can produce more than a megawatt of power, with new models producing more than 3 megawatts. Smaller-scale turbines are also being developed and may be better suited to on-site university applications. The turbines are also extremely reliable, with almost zero maintenance downtime. Wind power is an intermittent energy source, meaning that electricity is only produced when the wind blows, so wind power is not usually tied directly to a specific building, although a university in an area with significant wind may elect to install turbines on campus to help meet electricity needs. A new installation at the Massachusetts Maritime Academy is one such application.[7] As with PV technologies, a number of states offer programs to subsidize the installation of wind turbines.

Geothermal Geothermal systems use the earth to generate heat or cooling for buildings. A heat-exchange mechanism is used to tap the nearly constant temperature of the earth's surface (from 10 to 1,500 feet or more down). Ground-source or geothermal heat-pump systems consist of pipes buried in the ground near a building. In the winter, the warmth from the earth is used to heat water to around 55°F. The temperature is raised to 70°–80°F by a heat pump that in turn provides heat to the building. In the summer, hot air is pulled from the building and the relatively cooler temperature of the ground is used to cool air. Geothermal heat-pump systems are relatively cost effective and run cleanly. They do require some electricity to provide pumping energy. In some areas, deeper pockets of warmer geothermal energy are used to generate electricity or provide more comprehensive heating and cooling systems, but this application is rare.

Biofuels Biofuels or biomass includes trees, crops, and agricultural and forestry wastes that can be used to make fuels, chemicals, and electricity. Biomass is a domestic and renewable source of energy, although appropriate technology must accompany it for it to be "clean." Applications in universities and colleges usually include wood or wood-pellet boiler plants. For example, Mount Wachusett Community College recently installed a biomass heating plant to burn wood waste from the local furniture industry,[8] and the Society for the Protection of New Hampshire Forests uses a wood-pellet boiler to heat its offices. Eastern Connecticut State College runs one small boiler (100 horsepower) on B20 (20 percent biofuel).

A variety of fuels can be made from biomass resources, including the liquid fuels ethanol, methanol, and biodiesel, as well as gaseous fuels such as hydrogen and methane. Fuel such as biodiesel—a mix of biofuel and diesel—can be used in place of #2 heating oil or in diesel-powered vehicles. While the net benefits for climate change from these fuels vary, the benefit of using a local resource can be attractive. If the primary goal of switching to biofuel is a reduction in climate-altering gases, a careful life-cycle assessment should be conducted to ensure that the biofuel is consistent with the goal.

Integrated Approach

The most effective strategies for addressing campus energy use will include a combination of efficiency, fuels, and technology. The most successful energy approach will evaluate and implement a combination of appropriate technologies as part of the larger effort to reduce emissions. Reliance on one option alone will not solve the problem.

Decision Making for Climate Change

Many people make climate change–related decisions in university facilities. We introduced these actors in chapter 4, but expand on their roles as they relate to buildings here. Those who wish to affect climate change action on campus should take care to work with a range of people and to understand and appreciate their unique roles and their relationships to others. Building systems and energy delivery are complicated. Appreciating this complexity is critical to successful projects.

Administration

Trustees and high-level administrators make decisions that affect planning and resource allocations, which in turn can affect building locations, budgets, and sometimes actual design decisions. Trustees, overseers, and high-level administrators have varying degrees of input into decisions about buildings, but they set the tone and direction for planning, fundraising, and often aesthetics. Their institution's legacy, long-term plans, and academic and financial health are their concerns. Administrators and trustees play a critical role in goal setting, planning, and project budget decisions that have long-lasting consequences for buildings, energy systems, and climate change action. In addition, trustees can leverage other leadership roles in business, government, education, and other not-for-profits such as hospitals, cultural institutions, and faith-based organizations to influence climate change action on a larger regional scale.

Construction Personnel

The college or university construction department generally oversees new-building design and construction and major renovations of existing buildings. It is their job to see that budgets and schedules are met and that university standards are followed, and to act as day-to-day project managers. They also must understand and represent the diverse set of constituencies from within the institution. The interested parties may include the "client" department, maintenance, telecommunications, housing, regulatory requirements (such as zoning, stormwater management, historic restrictions, and the Americans with Disabilities Act), development office, community relations, donors, trustees, and possibly more. Some of the interests are in conflict. Construction managers must accurately represent these needs to the design and construction teams.

At Tufts the construction management function is an internal one, but at some state institutions this may be a department or an individual in operations or may be an agency that is outside of the university altogether. For example, the Massachusetts Division of Capital Asset Management oversees the design and construction of nonresidence buildings at the University of Massachusetts and the state's community colleges.

Construction departments or managers rarely design and build buildings themselves. Instead they rely on consulting design teams of architects and engineers and construction companies to design and build new buildings. Modern university buildings require dozens of specialists to properly design them. While the team is headed by an architect and an architectural firm that oversees the building shape and design, other members of the team include structural engineers to ensure that the building will withstand earthquakes, the weight of the building contents, and high winds; mechanical engineers who deal with airflow through the building, space heating and cooling, and water heating; electrical engineers who ensure that sufficient electrical power is provided to the building and its systems; civil and environmental engineers who design systems to handle stormwater, runoff, and grading; lighting engineers; plumbing engineers; landscape designers; and interior designers.

The design team is responsible for designing the building to meet the institution's needs. Each member of the team brings specialized expertise to the process and each can be a valuable source of information. Traditionally these design professionals compartmentalize their work, each playing a role. The interconnections between specialties are understood and the team works together. Yet the segmentation of work areas can lead to missed opportunities, and greater coordination can yield important benefits for achieving energy-efficient and high-performance buildings. Furthermore, close coordination between construction and facilities personnel ensures that buildings are designed and built in a manner that makes them more durable and easier to maintain.

Large renovations are also often the purview of construction personnel who oversee consulting design teams. In these projects, existing conditions are complicated and may not be well documented, especially in older buildings that have undergone many transformations. Close coordination between design, construction, and facilities teams will be essential in these renovations.

Facilities Personnel

Once the building has been constructed, the job of maintenance and problem solving falls to the college and university facilities staff of electricians, plumbers, mechanics, engineers, energy managers, and other tradespeople and their supervisors. Facilities personnel make many decisions with short- and long-term implications for reducing energy consumption, but these decision makers are generally also concerned about immediate costs and building function and/or occupant comfort. The facilities department is in the inevitable (and unenviable) position of having to balance occupant needs against the ability of building systems to meet these needs as well as available financial resources or expertise to fix immediate, chronic, and systemic problems in a timely and unobtrusive way. Facilities staff are often required to respond rapidly, quietly, and cost-effectively to complicated problems. They are expected to solve problems, even if building systems are inadequate or funding does not exist.

Members of the facilities staff who have detailed knowledge of energy systems include the energy manager, engineers, electricians, HVAC mechanics, and boiler operators. At Tufts, our most successful climate change action projects have resulted from the commitment and hard work of the facilities staff. It is critical that climate change advocates, interested faculty, and motivated students acknowledge the in-depth and pragmatic knowledge of their institution's facilities professionals. To be sure, like all of us, they have things to learn and may themselves be barriers to change, but their knowledge of existing systems and the problems that must be addressed by any proposed solution cannot be underestimated. We take students in a climate change course on a facilities tour to meet and talk directly with the people who operate the university's infrastructure, including the central heating plant. Students are amazed at the scale and complexity of the systems that the facilities department handles and are surprised that fuels, efficiency, and delivery of service are at the center of their decision making.

Many college and university facilities managers will tell you that they are short-staffed and lack sufficient financial resources to handle the day-to-day needs and regular maintenance required by the campus. They must often handle crises. One Tufts engineer describes these crises as the "undoing of the facilities zipper." In short, he says, the zipper "is not a term you can find in the facility handbook. The 'zipper' is the invisible thing that holds all the past screwups in place, capable of defying gravity and reversing time. When the zipper comes undone, illogical sequences of things that people cannot believe would ever happen start to happen with regularity. It, like the copier failing before the big meeting, happens only when you are least capable of dealing with it."

The facilities zipper is important for climate change advocates to understand and appreciate. New technologies or new methods, introduced in critical building systems without training or replacement parts, can undo the facilities zipper and create roadblocks for good projects for years to come.

Building Users

The expectations and habits of building users also play a role in the way that a building generates heat-trapping gases. In an occupied building, the addition of equipment, the opening and closing of doors, the scheduling of events, and the use of equipment when spaces are unoccupied can influence energy use. When new buildings are designed or renovations are planned, occupant desires (or expectations) can have important results for efficiency. For example, if a new music building's lobby is designated as a place where small groups will occasionally perform, the resulting lighting design might use theater-style stage lighting throughout the lobby when significantly more efficient alternatives exist for that type of space.

Depending on how a building is designed and operated, building occupants may control many uses of electricity and the level of heating or cooling (either directly with a thermostat or by opening or closing a window, or indirectly by complaining). Table 6.2 shows some of the unintended consequences of building-occupant actions. Those in charge of room scheduling, conferences, and special events influence the extent to which buildings and their energy-using systems are utilized and available for reduced loads.

Table 6.2
Consequences of building-occupant actions

Action	Consequence
Use of large numbers of computers	Overheating of space
Overheating space that contains thermostat for other spaces	Underheating of adjacent space
Leaving windows open	Inaccurate thermostat readings result in overheating or overcooling of other spaces
Use of space heaters	Increased electricity use
Leaving electrical equipment on overnight	Increased electricity use
Vacation and night use	Increased electricity use

Energy models and predictions of energy savings from retrofits depend heavily on predicting how the building will be used (hours and intensity). Some systems deliver savings when the building is not in use (e.g., occupancy controls or temperature setbacks) and others deliver savings in occupied mode (e.g., variable-speed motors, efficient chillers, and so on), but both require an accurate estimation of building use to be credible. At Tufts, despite our best efforts to predict energy use in a new Wildlife Clinic building, electricity use dramatically exceeded predictions. When we visited the building, we learned why—six large freezers lined the walls of the storage area. These units, and other electricity-using equipment critical to the services provided by the clinic, were not discussed when we modeled the building's electricity use. Careful attention to building users' equipment and actual use patterns should be gathered as part of the planning process and before undertaking any projects.

Climate Change Advocates in Facilities

To be effective, the advocate will need to partner with the right people, ask the right questions, evaluate the answers, and persevere. As described earlier, the advocate must understand how the institution works and work within systems if lasting solutions are to be developed. In the early days of the Tufts Climate Initiative, we found that it was effective to provide seed funding when we could for projects such as a photovoltaic installation, a solar hot water system, and third-party engineering. We also invested heavily in providing facilities personnel with access to information at conferences (paying for them to attend or bringing speakers and experts to campus). TCI also partnered with the director of facilities and the energy manager to raise issues to a high administrative level, coupling these issues with other university priorities such as education of students, funding, and visibility. The result has been a major shift in the way many decision makers in the facilities department think about energy, climate change action, and buildings themselves. This shift is resulting in increased investment, heightened awareness, and greener buildings.

We are constantly reminded that the climate change actions described more fully in the next chapter such as lighting efficiency and the installation of solar thermal systems are simply concepts. Making them a reality is highly institution and building specific and will require careful attention to the design and implementation that is right for the place. Simply because a strategy works on another seemingly similar campus is no guarantee that it will in work on yours.

As we discussed earlier, a climate change advocate is often a generalist, but buildings are complicated systems. Advocates who seek opportunities to reduce climate change impacts in buildings must partner with appropriate campus construction, planning, or facility managers, engineers, and energy personnel. Effective advocacy related to buildings will require this partnership. In addition, an effective advocate must become familiar with building terms and principles, but must recognize that general knowledge is not a substitute for the detailed engineering and expertise of architects and engineers. As we have worked on these issues at Tufts and talked to hundreds of faculty, staff, and students at institutions across the country, we have found that a failure to appreciate existing university staff—especially in the facilities and construction departments—and a failure to understand their professional objectives and time constraints under which they operate, as well as a failure to form productive partnerships with them, are the most common mistakes that slow progress and destroy an advocate's credibility.

We like to say that it is our job to ask the right questions and to evaluate the answers. Examples might include questions about the role of alternative fuels, equipment sizing, and design expectations. We have asked, "Do those south-facing windows need to be from floor to ceiling or can punched windows do the same job?" "Now that we have downsized the heating load, can we downsize the boiler too?" "Will anyone use this building during the hottest day of the year and if not, do we need to design systems that accommodate those worst-case conditions?" In asking questions and listening to answers, we hope to raise commonsense issues effectively. Too often we have found that common sense goes missing because building projects are so complex.

New Buildings

Our colleague Bill Moomaw has noted that "our institutions of higher education are like riders on a bicycle. Unless they are moving forward with new construction, they fear they will fall over."[9] Since most campuses are planning to grow rather than shrink their physical plant, efficiency and climate change action in new buildings are critical.

New construction creates the theoretical opportunity to "do it right the first time" from the perspective of energy efficiency and emission reductions. In practice it can be challenging to fully realize these opportunities in light of budget, schedule, limits to design expertise, and discomfort with new or unfamiliar systems. Meeting climate change goals is part of a larger effort to create "high-performance buildings"—buildings that have little impact on the environment or the occupants that use them. How a building uses materials and affects the surrounding landscape and the community are also important considerations. Taken together, this larger context for thinking about buildings, improving their design, and reducing their impact is a powerful motivation and legitimate investment opportunity for institutions that should have the future in mind. It is also an opportunity to educate future generations of students.

Many of us have had firsthand experience with building construction or renovation, whether at a personal level (new house, kitchen renovation, new school in the community) or in a professional capacity (renovations to campus buildings, or a new building for the department). A nearly universal outcome of building experiences is an appreciation of the complexity of the undertaking, and recognition of the many ways that problems can arise. Another near-universal outcome of construction projects is frustration with design and construction professionals, the prevalence of Murphy's law, and the near certainty that budgets will be overspent and/or schedules will slip. Rarely do even the simplest projects, from renovations to new construction, go right. Climate change action, and all the elements of high-performance buildings—durability, health, and minimized impacts on all aspects of the environment and the community—seem to add yet another layer of complexity when first introduced, yet in fact these efforts can often result in greater simplicity with smaller systems, fewer mechanical systems, and more durable construction.

Good Buildings, Green Buildings, and Climate Change Action

Any construction or renovation project should meet user needs and be completed on time and within budget—these are "good" or "effective" buildings and are challenging to build under any circumstances. Teams that build these "good" buildings are experienced, work well together, and take the time to develop sound plans and construction documents. There are few change orders (design changes during construction) and the inevitable conflicts are resolved quickly. In a good building the building owners and occupants are happy and the parties involved in design and construction make money. Good building teams will want to work together again.

Green buildings must first be "good" buildings. Otherwise they run the risk of simply having green elements in mediocre buildings and too often the green elements may be blamed for the buildings' shortcomings. Often, by asking the fundamental questions, especially early in the process, better buildings will result. Better buildings are those that serve the university needs more fully, especially over the long term, and tread more lightly on the planet. Climate change action can occur in buildings that may have problems, by taking incremental and piecemeal steps, such as by using a cleaner fuel or improving the efficiency of one building element. However, climate change action will be most successful, long lasting, and educational if it is integrated within a comprehensive design—one that connects building systems, rather than treating each element as a system unto itself. Climate change action also will be most comprehensive in a building that is a green building, achieving a variety of goals such as energy reduction, a high indoor air quality, and waste reduction.

Building-Project Goals

The greenest building you can build is one that you do not build. The unbuilt building does not use fuel, does not have impervious surfaces, and does not require that forests are harvested. Reusing old buildings, using existing buildings more fully, or doing without is nearly always the decision with the fewest environmental impacts. Although it is hard to fundraise for or to add new programs in rehabilitated existing buildings, the first important question that should be asked about a project is: Why do we need this building?

Once the determination has been made that the building must be built or the renovation undertaken, the team should ask itself "What results do we want?" Decision makers for the Oberlin College Environmental

Science Center clearly wanted to demonstrate state-of-the-art methods for buildings when they began that cutting-edge project.[10] Many other buildings will have a primary purpose other than this. Zero– or low–climate change impacts can be a goal—often one that will need translation with specific examples. While creating a building with no net impact on the environment is a laudable goal, it is often nearly impossible in the world in which most of us live and work, unless offsets are used (see chapter 8). Nonetheless, stating goals up front for building performance, particularly around energy use, can drive responsible, successful design decisions. When possible, the process is strengthened if the goals are quantitative and specific.

It is increasingly common for colleges and universities to include green building goals, such as those specified by the U.S. Green Building's Council's Leadership in Energy and Environmental Design (LEED) ranking system (see a more complete discussion in chapter 7). Stating these goals at the start can also help to motivate fundraising and inform all steps in the design and construction process. Several tools can facilitate the discussion. EPA provides an Internet-based tool that helps manage energy during the design of a new building. With Target Finder you can set an aggressive energy-performance target for a building design and compare your estimated energy consumption to the established target. Target Finder provides an energy-performance target rating for whole-building energy use.[11]

Decision Making During the discussion of a building's goals, it is helpful to discuss decision-making methods for design. For example, will all decisions be driven by the amount they cost up front? Or will life-cycle analysis—the cost of the decisions over time (see the discussion below)—be considered? If so, what criteria will be used? The roles of the client department, the facilities managers, and maintenance personnel in the design should also be discussed at the outset rather than ignoring their input until key decisions have been made. Many times, universities undertake major renovations and build new buildings using implicit, unstated decision-making methods. For instance, design teams may communicate only with the client (user) department, resulting in unrealistic expectations or designs that are not practical. By thinking carefully about project goals and establishing decision-making procedures at the outset, better buildings are more likely to result.

Building Budgets

The budget for designing and constructing a new building is often determined early in the process. It is determined by need, fundraising, and/or revenue streams from the building (e.g., a residence hall can generate revenue). In this way, the university can plan for the project and determine its scale. However, there are many factors that determine the budget that may not be known from the outset, such as the details of the building program (the types of activities the building will house), preliminary design conditions, project goals, and final designs. Sometimes there is pressure to present an optimistic budget in the early days in order to secure project approval. Over time, budget pressures increase as costs escalate (inflation) and new design and programming ideas are introduced. Careful management of design is critical to managing the budget to ensure that it is sufficient and to ensure that the proposed design is within budget. If the budget is too low, energy-efficiency measures can often be the first to suffer from the budget-cutting process (often called *value engineering*). However, not all universities and colleges proceed in this way. Some institutions and private-sector corporate clients determine the project budget only after significant programming and design are underway. These projects often use a collaborative project budgeting process known as *construction management at risk* to determine budgets along the way rather than with a bidding process at the end of the design phase. With construction management at risk, budgets may be more realistic and working budgets can be binding.

Approaching green building elements (and energy efficiency in particular) as a series of additional items that should be included in the project is nearly certain to add costs to the project. However, addressing the project systematically and linking all building components may have the opposite effect. For example, buying a more efficient air-conditioning unit may be more costly than a less efficient unit with the same cooling capacity. But taking steps to reduce the cooling load through design and engineering can result in a building that needs a smaller unit, so the smaller, more efficient unit may be less costly than the larger, conventional unit.

Life-Cycle Costing Life-cycle costing is the study of the total project cost—including construction, operation, and maintenance—over the projected life of the building. This analysis is typically applied to individual building elements; however, life-cycle costing can also apply to a collection of systems that, taken together, can improve building performance. Comprehensive thinking about life-cycle costing throughout the project, from its inception, is most effective and can save time. Cutting design-build budgets to the bone without considering the effect on building performance or operating and maintenance cost is far too common, especially in an era of rapidly escalating construction costs, but it will never produce good design.[12] Design teams need to be reminded that the university is the owner and occupant and will operate the building for decades, unlike some of their commercial developer clients. For example, a college can cut first costs by including an elaborate building entrance (which is visible to users) while purchasing a less efficient air-conditioning system (invisible to the building users); however, the operating costs of the inefficient air-conditioning system will be higher than necessary every year the building is used.

There is a saying that "we never have money to do it right the first time, but we have all the money to fix it." Clearly this is not an effective strategy from any vantage point; it does not conserve funds, time, or environmental resources. According to Steve Katona, then president of the College of the Atlantic, their building policy effectively requires a life-cycle approach. To begin construction on a new building, the college has to have in hand all of the funds for construction, operation, and maintenance.[13] We think this is brilliant and would like to see this policy in place at all institutions. With the decision-making tools now available for understanding energy scenarios, project decision makers can readily calculate the savings that will accrue during operations from initial investments in greater efficiency.

One of the key pieces of life-cycle costing is to project utility costs into the future, since higher future costs will create a faster payback for early investments. As we discussed earlier, this is a challenging task and one where the assumptions should be explicit and well understood. To be sure, the analysis should include the life-cycle costs at current prices, but scenarios should also be run with a variety of prices so that decision makers can understand the risks of increasing electricity and fuel costs.

At Tufts, we are still waiting for a crystal ball that can provide us with the most accurate estimates! Until it arrives, we must rely on using scenarios.

Design Fees Higher design fees alone will not guarantee better design. We have seen big-budget projects result in disintegrated, nonsustainable design. Design budgets typically lack incentives for engineers to add value. One remedy is to reward designers in part with shared cost savings from better design, such as smaller mechanical systems or improved energy efficiency; however this approach has not yet been embraced by the design industry.

Project Schedule and Timing

Because new building construction has a long lead time, often a facility is planned and even designed before an announcement is made to the campus community. As the design process moves from the general to the specific, it becomes more costly to make changes at each new stage. For that reason, student and faculty efforts to "green" buildings are often discouraging since these postannouncement ideas are often "too late" to inform the design process. Making changes in the final stages of design is a costly and complicated process, and furthermore, fundamental design opportunities such as building orientation, location, size, and scale have been determined long ago.

We would like to see the implementation of much more transparent project planning and design processes with frequent opportunities for community interaction so that student and faculty expertise can be incorporated. When faculty and student research relates to energy-efficient technologies, recycled building materials, and new ideas about landscaping, it is desirable to demonstrate these ideas in campus projects. But these demonstrations will require early collaborative involvement that may be outside the status quo.

TCI's involvement in the Wildlife Clinic, a new building for wildlife medicine, was late. We were brought into the process during the design phase, with the effect that some of our recommendations were accepted and others were not. In our postproject review of the Wildlife Clinic, the Tufts Climate Initiative concluded that we needed to develop a strategy in which the issue of climate change and the opportunities afforded by

Box 6.2
Questions to ask of a design team prior to hiring

1. What is the greenest project you've done?
2. Can we review the construction documents?
3. Can we see meeting minutes from that project?—use to assess how the team is thinking.
4. What strategies did you evaluate?
5. What strategies did you include? And why?
6. What was the most problematic building failure you've had in the past five years?
7. What did you do about it?
8. What would you expect a building such as the one we want to design to have as a heating and cooling load? What would you do to reduce that load?
9. Give us an example of how you considered the building as a system and made design decisions based on that thinking.
10. How much will your fee increase if we apply for LEED certification?

Source: Adapted from a talk by Marc Rosenbaum at Northeast Sustainable Energy Conference, March 13, 2003.

participate in these decisions and seek to influence them by pointing out design decisions with unintended costs for construction and operations.

If innovative technologies such as solar or geothermal are planned, it is critical that the professionals who design them are as carefully screened as any others are. For larger projects, it is especially important to determine that their expertise is consistent with the project's scale. Our experience at Tufts suggests that some design firms of specialty technologies, particularly solar technologies, are still learning how to adapt their designs and business practices from residential to institutional applications where formal responses to requests for proposals, liability insurance, strict code compliance, and coordination of construction documents with those of other designers are standard business practice. We urge these professionals to work hard to move solar and other technologies into the mainstream by becoming fluent in the existing ways of doing business.

At the same time, conventional engineering firms should develop capability in these specialties in order to facilitate the coordination among design subcontractors. Photovoltaic installations should be the domain of an electrical engineer and solar thermal should be the domain of the engineering firm designing the plumbing and hot water systems. In this way the process can be streamlined, saving money and time and finding creative applications.

In many cases, it is important to also assess the process for design that the design team is comfortable with. Will the design team willingly accept input from a third party if necessary? For example, it is often useful to have a third-party engineer or commissioning agent review plans and designs for opportunities for additional construction and operational savings. If the design team is hostile to this interaction, opportunities may be missed. It is also important to determine how possessive the architects feel about "their design"—will they refuse to make changes that the university client wants? For instance, will the team refuse to consider awnings for shading, photovoltaics, or light shelves because these alter the building aesthetics? Furthermore, will the design team encourage comprehensive discussion about the implications of design choices? As an example, if the university client believes that it wants a one-story-long building shaped like a snake, will the design team point out that a building of that design will likely use much more energy and usually cost more to construct than a more compact design?

The Design Process

The design of a new building is a process for moving from the general to the specific and has a number of well-defined steps. While green building advocates suggest that the traditional design steps (and the typical compartmentalization of design that tends to occur within those steps) is the largest barrier to good buildings, the climate change advocate needs to understand them nonetheless, since they are still widely practiced. Each step has a series of decisions that affect the final outcome. Here we provide a brief summary.

Programming Building programming is the process of defining the needs that the building will meet. Then the needs are translated into categories of spaces, space requirements, project goals, and preliminary budgets. Climate change action, energy action, and green design goals are critical to articulate in this phase of the work.

Schematic Design The schematic design phase is the preliminary design phase of the project. In this phase the building is located on the site, its size and shape are determined, and the general layout is established. The early design work is where the most opportunities for efficiency are gained or lost, so it is critical to think carefully about design decisions early in the process. Again, budgeting is often a part of this phase.

In a conventional design, much of the work in the schematic design phase is driven by the architect, with the engineers participating to discuss what is needed to make the building stand up (structural) and be comfortable (mechanical). However, this early design phase is a logical place to think about the building as a system of interconnected, rather than separate, design pieces. A design charrette (a meeting of the team to discuss the building and its goals, and to create ways to meet the goals) can be instrumental for bridging this gap at this early phase of the process. But the project manager and the team will need to work hard throughout to overcome what is usually a linear design process "in which the differing disciplines each do their piece of the work in isolation and then hand the design on to the next specialist. This compartmentalization is hard to overcome, and undermines both design integration and building performance."[14] Early design efforts are likely to get the best results if the team works together collaboratively.

Design Development Design development is the phase when the design is fleshed out and many more of the building details are determined. As designs and ideas are "test fit," there are often changes. At this stage, traditionally the engineers begin to develop their design; however, they often rely heavily on past designs and standard rules of thumb, rather than on finding creative ways to solve problems unique to a particular building. Often the engineers ask to be involved during this phase of the project or later, but at this point many opportunities have been lost. Once a basic design has been shaped, it sometimes customary to engage a general contractor as part of the team to find cost-effective construction options as well.

As we have discussed earlier, an integrated design approach can help ensure that the building is designed as a system. Systems thinking needs to be combined with an integrated design team approach that helps to overcome the compartmentalized design solutions to problems. Buildings should be outcomes of a strategy to meet objectives, rather than an embodiment of individual design elements that are pieced together. Amory Lovins of the Rocky Mountain Institute says, "If you just optimize a component like thermal insulation by itself, you 'pessimize' the building by not doing a life cycle comparison that counts all the system's capital and operating costs."[15] Lovins also likes to say that multiple components interact in ways that are not obvious when you are looking at them separately. Energy systems are the most likely to benefit from systems thinking—often they can be smaller and simpler if the rest of the building is designed with low-energy building goals in mind.

Building energy models, daylighting models, and other simulations usually begin during design development. Energy modeling is critical for determining the trade-offs in the building—whether or not there are opportunities to improve the building envelope and downsize the mechanical systems, for example. Energy-performance modeling is now a standard tool of many design firms that include architects and engineers.

Value Engineering During design development, cost estimates are developed (some buildings have preliminary cost estimates during schematic design as well). Cost cutting can often occur during this phase, and often the process of "value engineering" starts. This process can frequently be detrimental to energy measures, especially those that are incremental rather than integrated design changes. However, a value-engineering process can be strengthened if it looks beyond first cost and beyond the cost of single measures to include operating costs, life-cycle costs, and the interconnections of systems.

Construction Documents Construction documents are the building plans and corresponding narrative specifications that describe how every part of the building will function and how these parts will fit together. During the preparation of the construction documents many details of the building are finalized; however, the general design parameters are usually established much earlier. Nonetheless, there are many opportunities to affect the specific technologies selected, if not the larger design, during this phase of the project. For example, the types of lighting or location and type of lighting controls will be determined in this project phase. The chiller, boiler, and ventilation design will also be finalized.

The construction documents fully describe everything from the building's goals to specific products, installation techniques, and other methods. These documents should conform with any college or university construction standards or requirements (see chapter 7).

Bidding and Selection of Contractors Once the construction documents are finalized, the university will typically receive final bids from contractors, although some institutions work more closely with a contractor through the design to establish the budget along the way and ensure the building's constructability early in the process. Especially in public institutions, the bids are limited to the exact contents of the documents. In some institutions, particularly private institutions, there may be earlier involvement of a contractor during the process. The goal is to create more credible bids and to link the design and construction more closely. This can be beneficial for ensuring cost-effective design and for identifying alternative construction techniques to better meet building and budget goals.

The bidding documents and contract documents should also include specific reference to Leadership in Energy and Environmental Design or LEED (see chapter 7), if LEED is a criterion, as well as to energy efficiency. Even the advertisement for bidders and the invitation to the bidders' conference should reflect that intention. The bidders' meetings should explain the LEED requirements and the energy-efficiency goals, both in the instructions and in the meeting. Once a winning bid is accepted, the contract is awarded. In the contract, LEED requirements should be incorporated into the supplementary conditions and linked into the contract.

Construction

Construction of the building according to the plans is perhaps the most complex step. In large and small buildings there are inevitably problems. However, in good buildings and collaborative processes, these problems are solved effectively and efficiently.

Depending on the campus culture, community awareness about a new building may be low until construction begins. At that time, there can sometimes be a flurry of interest in affecting changes in the design as students and faculty visualize the effect of the new building on them and the campus environment. At this point it is too late to accommodate changes. The key to avoiding disappointment among community members is to communicate the construction plans very early in the process.

During construction, any changes can be costly. These "change orders" or departures from the plan can result from the owner changing its mind, errors in the construction documents, new technology, budget tightening, or a host of other factors. Good projects usually minimize change orders and can save substantial costs by getting it right the first time. Seldom is it cheaper to do it right on the second try.

Regardless of the contractor, the construction process needs careful supervision by the design firm and by an owner's representative (project manager) in order to prevent mistakes or to catch them before it becomes costly to change them. The substitution of products is an important area to pay attention to during the building process. The sealing of exterior spaces and the installation of all ductwork are two of the many procedures that should receive careful scrutiny because of their climate change implications.

Contract Closeout

The contract closeout will require full commissioning (see chapter 7) of all systems. In addition, closeout typically includes a building warranty period—ideally of at least one year—to determine that the building and its systems actually work as designed.

Design-Build Projects

Design-build projects use the same firm to both design and construct a building. Often this strategy is used to reduce costs and shorten the construction schedule. Successful design-build projects require that clients, in this case the university, specify the characteristics of the building in great detail in the request for proposals (RFP). In fact, the design-build process is essentially the antithesis of achieving the LEED certification objectives. In a design-build process the developer is usually looking for financial advantage rather than improving performance objectives. However, the design-build process can be crafted to achieve desired goals by carefully articulating the objectives the building must achieve in the RFP. The Connecticut Department of Public Works Facilities Management made LEED a part of the design-build RFP for three Connecticut State University system residence halls at Eastern Connecticut State University. Although the project had a challenging design process, the result is three buildings that will be LEED certified.[16]

A Case for Action in Existing Buildings

Many campus stewardship programs and environmental efforts worldwide have demonstrated far-reaching action in state-of-the-art new buildings that generate power or are highly efficient or both. Nonetheless, regardless of their features, efficiency, power generation, or on-site waste treatment, new buildings increase net resource use.

At universities there is the potential to replace old inefficient buildings with new. However, even in the case where a new building replaces an existing building, careful analysis is likely to show a net increase in resources in the short run, largely because new buildings generally add program space. Although there may be a benefit in the long run by replacing a very inefficient building with a highly efficient one, this is the rare case. In many campus situations, the old building remains along with the new.

Recently, the environmental movement has had a visible focus on new high-performance buildings. However, if a climate action effort concentrates only on new buildings without addressing its existing buildings, it does not address the need for net reductions in climate-altering gas emissions. This is like dieting by adding a low-fat food without adjusting the other foods in your diet. The long-term focus must remain on reducing the energy intensity of existing buildings, and many campuses have been working hard at just that task for decades.

In many existing buildings, climate change action will be invisible to building occupants, but may have significant benefit to reduction of greenhouse gases and to the bottom line. Many of the actions require changing hardware such as boilers, chillers, lights, and controls. Once installed and operational, these devices will continue to save energy over time. Unlike climate change actions that require members of the university community to undertake some individual action,

7

Tackling Emissions at the Source: Climate Actions in Buildings and Central Facilities

New and existing buildings are full of opportunities to make significant emission reductions. In this chapter we provide details on actions that can reduce emissions from campus buildings—actions that often apply to both renovations and new construction. We also touch on the opportunities to think about climate change action in central facilities that provide central heating, cooling, and electricity generation.

There are opportunities for reducing climate change impacts in almost every building. Identifying these opportunities can be easy or difficult, but carrying out the measures can be complicated and requires careful planning as well as thorough implementation and follow-through. Unfortunately, many building energy projects are not discrete, simple projects, but require complex engineering assessments and significant capital or systems analysis. It is very important for operations personnel to be part of any serious efficiency discussion as well as full participants in the design of new construction.

While every energy and climate change project has a degree of complexity, those that involve discrete systems are the simplest. The most common example is the replacement of conventional incandescent lighting with energy-efficient lighting and lighting controls. Other examples include replacement of aging motors with premium efficiency motors or fixing leaking steam traps. Projects that save electricity lend themselves easily to calculating savings and isolating into discrete projects. Some electric utilities will pay for project engineering or pay an incentive to encourage project implementation with demand-side management programs. Many of these electricity-saving projects also have rapid (two- to three-year) paybacks and can be undertaken piecemeal or comprehensively if funding is available.

Finding Energy-Saving Opportunities

In some buildings or sections of buildings there may be obvious signs of waste. Simple observations may tell you that the lights are on around the clock or that windows are open in the middle of a winter freeze. Conversations with building users may also tell you that the water is always too hot or that the occupants need to supplement heating or cooling with space heaters or window air conditioners. In many cases, however, comprehensive energy assessments conducted by a trained mechanical and/or electrical engineer are needed to identify the complicated set of existing conditions, building uses, design conditions, and actual systems in order to design appropriate and effective efficiency measures. These energy audits form the basis for planning and funding an in-depth energy-efficiency program.

In addition to learning about the physical systems for controlling and delivering energy in a building, talking with building users should always be included as part of any energy assessment. For example, a lighting program might install sensors to turn off lights when spaces are unoccupied, but such an effort might disturb a laboratory studying how light affects sleep in mice. Similarly, efforts to shut down buildings and reduce heating or cooling during weekends or vacations might disrupt necessary stable laboratory conditions used to measure the strength or curing rate of concrete mixtures in a structural engineering lab, or could interfere with grow rates of a cultured bacteria in a microbiology lab.

At the same time, the needs of building users should be evaluated in a policy effort to determine whether the university should meet all faculty and student demands. For example, if a large faculty office building can be "closed" during the summer weekends, significant energy savings will result due to reduced air-conditioning loads. However, if one faculty member expects cool summer temperatures in his or her office on a summer Sunday afternoon, the savings will quickly evaporate. Chapter 9 addresses policies regarding building temperatures.

Climate change advocates can be useful in identifying energy-saving opportunities and encouraging people to develop nontraditional solutions to energy problems. In addition, an advocate can help to bridge the gap between nonuniversity "energy experts" and university personnel or to raise interdisciplinary considerations such as the use of landscape trees for shading to reduce cooling loads. Tufts Climate Initiative staff frequently poll other institutions to find out how they handle situations to give credibility to new ideas or to identify creative and cost-effective solutions. When new technologies are proposed, TCI staff often track down others using similar equipment and validate manufacturer or engineer claims of reliability and ease of maintenance.

Specific Opportunities for Reducing Emissions in Buildings

A range of opportunities exist for reducing energy use in buildings. These may be tackled piecemeal or comprehensively as budgets and time allow. In new construction, a target goal to make a building 30 percent, 50 percent, or 70 percent more efficient than required by building code is a critical step for maximizing design creativity on these problems. The range of opportunities is large. This section highlights several specific opportunities for getting started in addressing climate change actions in campus buildings.

Building Envelope

The building envelope is the building's outer skin—windows, doors, walls, and roof. Insulation provides the greatest payback when included in a new building, but when carefully done in an existing structure, it can provide significant value to the owner. Areas for attention are the roof, attics, walls, soffits, and windows. Reducing air infiltration and leakage can help to control the indoor environment and to allow deliberate and efficient ways to provide fresh air into the building. While we usually think of savings opportunities from tightening the building envelope most commonly in residential heating in northern climates, the savings can also accrue, sometimes more significantly, with air-conditioning, especially in commercial and institutional buildings and those in warmer climates.

Windows generally transfer more heat than walls. They are also more expensive to build than walls. And while properly designed windows can help to reduce lighting needs, if done wrong, they can create glare and require complicated (and expensive shading). The wrong windows can allow large amounts of unwanted solar gain into a room. This problem is especially important throughout the southern United States, although in northern climates during winter the solar gain can be an asset. Careful attention to windows in new buildings and in existing buildings should address the solar gain and the associated increased cooling or decreased heating needs. Selecting windows that are the correct size and with the right light-gain and heat-loss characteristics is important in both new construction and renovations. Window selection should include attention to the glazing, the sash, and the frame as well as to durability and functionality. Similar considerations should apply to doors as well.

Minimizing penetrations, reducing or eliminating thermal bridges (ways for heat to transfer from the interior to the exterior), reducing glazing, and providing careful installation are all important opportunities. We are disappointed when we see a new building full of floor-to-ceiling windows that always have their shades pulled to avoid the glare. These buildings look wonderful on the drawing board, but they actually are more expensive to build and operate and provide less daylight than buildings with smaller windows that are located strategically and have glazing to reduce the glare. Architects should be well versed in window options and trade-offs and be prepared to discuss them with their university clients. Careful writing of specifications for building envelopes in new buildings and careful installation (overseen by an owner's representative) are critical for maximizing the opportunities for envelope efficiency.

The savings from building-envelope improvements can be substantial. In the Vermont Law School's Oakes Hall, envelope improvements, while at a premium cost, eliminated the need for perimeter heating systems—a dramatic first-cost and operating-cost savings.[1]

Light-colored roofs are another opportunity for reducing summer cooling loads in a building, since the light surface reflects heat rather than absorbing it. In buildings with a significant heating load, and little cooling load, however, there may be no benefit from a white roof since the incoming air is actually preheated by a dark roof under some conditions. This is another example of the complex and situation-specific nature of these decisions.

Lighting

Lighting can account for anywhere from 20 to 50 percent of a building's total energy use, so it is important to concentrate on efficiency throughout campus operations.[2] We know from research and personal experience that human beings are very responsive to different types of light. Artful lighting in a theater production can transform the audience's experience, and insufficient daylight is associated with seasonal affective disorder. Lighting can change the tone of a room; dining by candlelight seems to improve the taste of the food and the appearance of the diners. While recognizing the impacts that specialized lighting can have, the fact remains that most lighting at institutions is utilitarian, and improving its efficiency will reduce greenhouse gas emissions associated with electricity. An additional benefit of efficient lighting is that it can reduce a building's cooling load, thereby increasing its energy savings in buildings with large air-conditioning needs.

General Lighting Although lighting appears to be a simple concept, it requires expertise and instrumentation to select appropriate technology and to ensure that light levels are adequate. Lighting renovation projects involve more than just replacing a bulb. New technologies have emerged in the last decade that change how light is delivered. Lamps, ballasts, reflectors, and indirect and direct sources are among the many technologies that influence how light is delivered. Wall color, surface brightness, and daylighting are other considerations.

In new buildings, energy advocates should communicate lighting-efficiency goals early in the design process and push hard to get the lowest lighting power density (watts per square foot) and the most efficient fixtures. Architects frequently use decorative lights to showcase architectural features. These lights should be considered carefully and eliminated where possible. Whenever feasible, lighting and lighting controls that qualify for utility rebates should be specified.

In general university spaces, fluorescent lighting is better than incandescent. (See box 7.1 for types of indoor lighting.) The selection of the entire fixture, including lamp and reflector, is important for maximizing efficiency. Layout and switching can also increase opportunities for efficiency—for example, banks of lights along windows may be shut off while interior lights stay on, or lights in library stacks can switch on and off with occupancy controls if they are aligned parallel to the stacks. Lighting selection should also take into consideration the types (including lighting color and length) of lamps that the institution currently

Box 7.1
Types of indoor lighting

Regular Incandescent Bulbs
These bulbs, which provide most home lighting, are used in products from nightlights to floodlights. The most common incandescent is a pear-shaped bulb with a medium-sized screw-type base. Incandescent bulbs use electricity to heat a filament until it glows white hot, producing light. About 90 percent of the electricity used by incandescent bulbs is lost as heat.

Incandescent Spotlights and Floodlights
The reflective coating on these bulbs helps direct and focus the light. Commonly known as spotlights or floodlights, these bulbs often are used in recessed ceiling fixtures or outdoors.

Halogen Bulbs
Sometimes referred to as "tungsten-halogen filament incandescent bulbs," these bulbs contain a small capsule filled with halogen gas, which emits a bright white light. Halogen bulbs produce more light, use less energy, and last longer than standard incandescent bulbs of the same wattage, but they cost more and are hotter.

General-Service Fluorescent Bulbs
These bulbs are more energy efficient than incandescent bulbs because they do not produce heat. They are the thin, long tubes often used in kitchens for under-cabinet lighting, and in garages, workshops, and basements. The tubes can last from 10,000 to 20,000 hours—10 to 20 times longer than incandescent bulbs.

Compact Fluorescent Bulbs
These bulbs provide as much light as regular incandescent bulbs while using just one-fourth the energy. For example, a 15-watt compact fluorescent bulb yields the same amount of light as a 60-watt incandescent bulb. Compact flourescent bulbs last about 10,000 hours—10 times longer than incandescent bulbs.

Source: Federal Trade Commission, "Energy Efficient Light Bulbs: A Bright Idea," http://www.ftc.gov/bcp/conline/pubs/products/ffclight.htm.

stocks. In a recent Tufts project the architect proposed a three-foot version of a standard lamp that we typically stock in two- and four-foot models. A simple plan change was all that was needed to save years of headache.

Task Lights Task lights provide targeted light at a desk or specific work surface. Where overhead fluorescent lighting in classrooms and offices is outdated, inefficient, or poorly maintained, supplemental desk, table, and floor lights abound, often with incandescent or halogen lamps. Thus lighting projects should address the hardwired overhead lights, their fixtures, and the fixture placement to reduce this need, but should also address the softwired task lights. In some cases, such as office cubicles, task lights can provide greater light quality and reduce the lighting levels and associated energy needed for overhead lighting. Although this is rarely done in practice, task lighting and overhead lighting should be considered together as a single system. Tufts facilities personnel have found that upgrading the overhead lighting in residence-hall rooms to a brighter light has decreased the need for more energy-intensive desk lamps. Where task lighting remains, incandescent lightbulbs should be replaced with screw-in compact fluorescent bulbs.

Decorative table lamps are often used in campus spaces such as waiting areas for administrative offices. We have replaced incandescent lightbulbs in task and table lamps throughout campus, most prominently throughout the president's office suite. To date, TCI has given away over 3,000 compact fluorescent lightbulbs on campus to replace traditional incandescent lights. The only price we charge is receipt of the incandescent bulb in return for the compact fluorescent. We also promise a satisfaction-guaranteed or "incandescent lightbulb back" policy—but so far, we have had no returns.

Halogen torchieres (up lights on pedestals) are banned at Tufts and many campuses due to their risk of starting fires as well as their very high energy use. These bans are common among colleges and universities even though students like the diffuse light directed toward the ceiling. On some campuses, decision makers calculated that it was cheaper to replace the halogen units with compact fluorescent torchieres rather than pay for the electricity to run the lights.[3]

Accent Lighting Accent lighting is popular with architects to showcase artwork or architectural features, distinguish building elements or spaces, and add accent. Accent lighting is also commonly found in on-campus retail-type spaces such as bookstores and dining facilities. Wherever possible, halogen and incandescent lamps should be reduced or eliminated in favor of more efficient and longer-lasting alternatives such as fluorescents (new color qualities make these better choices than just a few years ago) and even LEDs (light-emitting diodes).

Hidden Opportunities In existing buildings, we have discovered a number of areas for quick lighting changes that can save money and have multiple benefits:

- Chandeliers are one example where switching to longer-lasting compact fluorescent lighting reduces maintenance costs.
- In dining areas, fluorescent lighting over salad bars and in refrigerators reduces heat gain.
- Gymnasiums and high-bay spaces can be lit effectively with high-intensity compact fluorescent fixtures.
- Outdoor lights last longer if they are compact fluorescent or other fluorescent technology.
- In addition, lighting can be improved by using lighter colors on walls and ceilings.

The maintenance benefits of efficient lighting (longer-lasting lamps) is another feature, and hard-to-change lamps should be targeted first in a lighting retrofit program.

Lighting Controls Even the most efficient lights use electricity when they are on and save it when they are off. It is not uncommon to see classroom buildings that are fully lit in the middle of the night or offices and classrooms that are well lit but unoccupied. College and university settings are good places to use lighting controls to save electricity, and new buildings should use lighting controls as a matter of course. Occupancy sensors will shut off lights when spaces are not occupied. Photosensors can shut off lamps when natural light provides sufficient levels. At Tufts we have implemented comprehensive installation of occupancy sensors triggered by motion and infrared energy to shut off lights in unused classrooms and offices throughout the university. We estimate that the energy saved by this measure will pay back the initial $600,000 investment in two to three years. Comprehensive installation of the technology allows for standardization to improve maintenance and troubleshooting. Nonetheless, lighting experts must assist with a project such as this to determine proper placement of the sensors and to ensure that the appropriate technology is selected. Organizations that installed relatively unsophisticated sensors in the 1970s may have their share of horror stories (e.g., motion-actuated sensors causing occupied rooms to plunge into total darkness), and we hope that those who feel burned by earlier efforts will try again with improved technology and expert advice.

Reducing light levels when light is not needed is a second type of lighting control that is becoming increasingly common and can save energy. These controls can dim fluorescent lights if the light fixtures include dimming ballasts. Alternatively, it can shut off just some of the fixtures. In some cases gradual dimming is needed, but in other cases it is more cost effective to simply select "two-stage" ballasts, or ballasts that allow the light to provide low power and full power. The control determining whether the light is on high or low power is provided by a photosensor that senses other light (such as sunlight) and powers the lights down, or by an occupancy sensor.

We have been experimenting with a second type of occupancy lighting control in spaces that must remain lighted according to building codes (e.g., hallways in residence halls), where we combine a two-stage dimming technology with occupancy controls to reduce light levels in unoccupied mode while leaving lights on to meet code.

Daylighting Daylighting is a technique that allows for maximum usage of sunlight for interior lighting needs. To maximize the advantage daylighting can provide, it is important that it be a goal of the design process and that lighting controls switch off lights or switch them to lower power when sunlight is sufficient. Daylighting can be as simple as properly placing windows to take full advantage of outside light, or as complicated as systems of light shelves that reflect light deeper into a room to increase the benefit from natural light. When properly integrated with lighting controls, daylight can help reduce lighting costs for building owners. In existing buildings, occupants can raise the shades and turn off lights when daylight is sufficient.

Motors

Comprehensive motor replacement should be part of any climate change or energy-saving effort. Electric motors consume 23 percent of the electricity produced in the United States. Over its life, electricity is 97 percent of a motor's cost.[4] In mechanical rooms, ceilings, elevators, vent hoods, and mechanical and plumbing equipment, motors move air and water throughout most institutional buildings. Other energy-intensive applications such as pumps, fans, and standby equipment are other areas for efficiency. Some motors run continuously, while others cycle off and on. Replacing motors as they fail with highly efficient motors can save enormous amounts of electricity. In many cases, it is even cost effective to replace motors prior to failure (with significant hours of remaining life) since the savings from reduced energy use are so significant. In some cases, motor-replacement projects can pay back in less than a year. If HVAC maintenance services are contracted, it is critical that premium efficiency motors are specified as replacement parts on a routine basis, to avoid getting the least expensive (first-cost) motor that costs a great deal more to operate over its life.

Mechanical Systems

Mechanical systems are often the largest users of energy in a building. The greatest opportunity for energy saving is to "rightsize" during new construction, but existing buildings also provide many opportunities for improvement because oversizing mechanical systems is very common. In the design of new systems and the renovation of old, it is important to discuss the operating assumptions, look for ways to reduce the hours of equipment operation, assess outside air intake and building pressurization, create single-zone systems, assess reheat systems (air is generally cooled and then reheated before it is delivered to a space in order to maximize comfort), use variable-air-volume systems, and more.[5]

Despite the effort to develop an integrated building design, it is still critical to pay attention to the efficiency of the mechanical (and electrical) systems that are proposed for any given building or renovation in order to maximize efficiency. Having the university engineer involved in any design work and/or hiring a commissioning agent during any mechanical design to act as the owner's representative will be well worth the money. System efficiency rating, controls, design conditions, and loading of equipment are all factors that need careful engineering and are outside the scope of this book.

In new construction and renovation, design teams should be asking "What is the most efficient equipment that can do this job and how much does it cost, both in first cost and operating costs?" Unless those questions are asked during design meetings, they are overlooked. Ironically, in design, the client is often given a wide range of choices about things such as flooring material or wall coloring—choices with little or no payback—but is given fewer choices about selection of the boilers or a detailed look at the design assumptions under which equipment is sized.

In addition to the efficiency of particular systems, it is important to evaluate alternative systems that may operate quite differently, but achieve the same result. For example, radiant heating (heating the objects, rather than the air) operates more efficiently in many instances and may be a good choice, especially where spaces are occupied continuously. Nearly a decade ago, the Union of Concerned Scientists installed heat pumps in addition to a very small high-efficiency boiler in their Cambridge, Massachusetts, office renovation; these continue to operate today. Solar thermal systems and geothermal heat pumps are other examples. As a climate change advocate, you can play an important role in identifying these opportunities and evaluating the benefits of technology transfer from a comparable application to your own institution.

Rightsizing "Rightsizing" or determining the correct size of needed equipment can actually save both first costs and operating costs. Systems are typically oversized—creating a comfort margin or expansion possibilities that are not needed. Larger systems cost more at the outset and may run less efficiently when they operate at low loads in order to meet the real conditions.

Rightsizing can also have another dimension, by determining that adequate ventilation is provided to spaces, but only when needed. In one experience at Tufts, we found that the consulting mechanical engineer planned to provide fifteen air changes an hour throughout an entire animal clinic building. This is a high level of air changes (a complete change every four minutes!) and was required, but only in spaces occupied by animals; it was not required in corridors, storage rooms, and offices. By reducing the areas with such extensive fresh air requirements, we were able to purchase smaller, more efficient systems at a lower cost than originally proposed.

Heat Exchange When air is exhausted from a building there is an opportunity to use it to preheat or precool incoming air using a heat exchanger, enthalpy wheel, or air-to-air exchanger. The Tufts Wildlife Clinic installed a heat exchanger that demonstrated a five-year payback. This technology is especially cost effective for spaces or buildings with high rates of air changes.

Consolidation of Systems Because many university buildings are relatively old, renovations can create opportunities for significant emission reductions that if missed can be sources of significant increase in emissions as new uses are added to old buildings. There is a temptation to add small systems bit by bit to accommodate the specific renovated area, and these systems are rarely efficient. On our medical campus, one building was converted piecemeal from a garment factory to a high-tech research environment. In 1999, the university spent $12 million to consolidate the nearly eighty HVAC systems into a single system. The project is being paid for out of energy savings over fifteen years. But if the building stock on campus is recent and was constructed to meet the current needs and with energy efficiency in mind, dramatic consolidations of this type may not be needed.

Steam Traps A steam trap is a compact, relatively low-cost automatic valve for releasing condensate (condensed steam) and noncondensable gases, and for preventing the escape of useful steam from a distribution system. Steam traps are designed to maintain steam energy efficiency while performing specific tasks such as heating a building. "Once steam has transferred Btus and becomes hot water, it is removed by the trap from the steam side as condensate and either returned to the boiler via condensate return lines or discharged to the atmosphere (a wasteful practice)."[6] Steam traps are an important element in efficient facility operations and in energy conservation. Neglected traps can waste steam since they generally fail in the open (or most wasteful position), costing many times the price of an effective inspection and maintenance program. At Tufts, the facilities department undertook a comprehensive steam-trap replacement project to improve the efficiency of steam distribution within buildings. We expect a project payback of less than five years.

Energy Management and Measurement

An energy management system (EMS) controls how energy is used and how building equipment operates. Measurement and monitoring of buildings and building systems with an energy management system is a critical tool for successful management of complex university facilities, tracking trends, comparing buildings, identifying problems, and providing efficient building management. EMSs may be sophisticated or simple. Sophisticated systems can be used to regulate temperature, increase or decrease fresh air, identify problems, and monitor conditions. Systems may also change the temperature to which hot water is heated based on the outside temperature (lower temperature when outside temperatures are higher). Energy management systems alone are not necessarily a guarantee of energy efficiency, but they are a tool for managing energy on a campus and in a diverse set of buildings. In general, EMSs save and manage energy by controlling equipment so that

- Equipment is running only when necessary.
- Equipment is operating at the minimum capacity required.
- Peak electric demand is minimized.[7]

EMSs also may be used to save energy by monitoring equipment and operational data, as well as for assessing a building's problems, energy trends, or efficiency opportunities.

Measurement and control systems can also be very simple and often these simple systems can be effective. For example, rather than operate ventilation controls from central facilities, a conference room can have user-activated on-off switches that turn on ventilation when activated and turn it off on a timer. In other cases, occupancy sensors can effectively turn off systems or lights when a room is not in use, lower fumehood sash heights based on occupancy, or reduce or increase air flow based on monitored levels of carbon dioxide (e.g., in a lecture hall) or carbon monoxide (e.g., in an underground parking structure). Our experience indicates that any user controls must be within acceptable ranges (and the EMS can provide those limits) and accompanied by very clear signage.

Scheduling EMSs are very effective for improving energy efficiency by improving the scheduling of equipment—turning it off when not needed depending on the time, day, usage, or outside conditions. Every piece of equipment that can be shut off twelve hours a day (rather than operating continuously) cuts its energy use in half. Periodic review of the scheduling sequences offers savings opportunities.

Feedback Systems Monitoring systems can also provide building users with feedback—an important element for raising awareness and encouraging personal action. Simple systems, such as that employed at SUNY Buffalo, use monthly paper charts to track energy trends in a given building. These charts are posted in each building. More sophisticated systems, such as that used in the Oberlin College Lewis Center, have real-time monitoring and web-based reporting of building conditions. In our new solar residence hall we have real-time displays to show electricity generation from photovoltaics and real-time building electricity and heating use.

Monitoring for Diagnostics Building diagnostics experts and astute facilities personnel will also use portable monitoring devices such as thermometers, smokers, and humidity sensors to detect problems and assess existing conditions.

Hot Water

Heating domestic hot water is energy intensive. There are three strategies for reducing energy used to heat water. The first is to decrease the use of hot water through water conservation measures. The second is to decrease energy use by insulating pipes, setting back hot water temperatures, and storing hot water more efficiently. For example, instantaneous water heaters will heat water only when needed. The third strategy is to use alternative fuels such as solar thermal systems or other heat-exchange systems to preheat water (see chapter 6).

Laboratories

Laboratory buildings are generally the most energy-intensive buildings on a campus. Laboratory fume hoods are large energy users since they usually run constantly, exhausting conditioned air (heated or cooled) from a room that must be replaced with additional conditioned air. Careful attention to their design is essential for any effort to control energy use and costs. The Labs21 program offers universities tools for addressing labs.[8] There are numerous options available to laboratory facilities to increase efficiency.[9] Some of these can be included or addressed in renovations to existing laboratories.

Attention to Makeup Air Because hoods exhaust so much air, a large volume of additional air must be brought in to replace it. This is called "makeup" air. One strategy is to use outside air that is directly pumped into the fume hood; this reduces the amount of air that needs to be conditioned. Another example is to make sure that controls are linked to the makeup air system. Some strategies can provide significant increases in efficiency, but depending on the college's climate, may cause discomfort for the hood user because the air is unconditioned (neither heated nor cooled).[10]

Fume Hood and Fume-Hood Controls Some hoods exhaust air at a constant volume, and others have occupancy sensors to lower the sash and reduce airflow during unoccupied times. In some applications the constant, low-volume hoods may save energy, and in others controlling the flow of air based on the height of the fume-hood sash can have a substantial impact on the efficiency of the fume hood. For example, some controls automatically monitor the height of sash and regulate the airflow accordingly. Additionally, some controls reduce airflow if there are no operators present for an extended period of time. These controls do have shortcomings in that their maintenance can be costly, and if several hoods are in use at once, it can overwhelm the building's HVAC systems.[11]

Rightsizing Selection of mechanical equipment and heat-recovery equipment to determine it is the correct size and is capturing waste heat or cooling from exhaust air is critical in labs. In addition, it is important to make the correct assumptions about how many fume hoods will be in use at the same time, a phenomenon called "diversity" (since a lower number of concurrent uses can result in smaller and more efficient equipment).

Heat Exchangers Heat exchangers capture the heat that escapes through the fume hood and use it in the building systems to increase the efficiency of HVAC operations. If some of the heat that is exhausted by hoods on a cold day is captured in a heat exchanger, it can be used in the makeup air to reduce the amount of heating the makeup air requires.

Lab Management and Maintenance Simply training users to shut the sashes when not in use can create significant benefits. In addition, ensuring that the maintenance staff has the tools and knowledge to maintain the equipment and to use controls to operate the space effectively can save energy.

Plug Loads Internal heat loads from laboratory equipment can be significant since lab spaces may also have large plug loads; these plug loads can vary substantially from time to time. Working with lab managers to properly manage these plug loads is a no-cost or low-cost way to reduce electricity usage.

Dining Facilities

Kitchens and dining facilities are more energy intensive than offices. Like laboratories, kitchens require extensive exhaust systems for the health and comfort of occupants, and conditioning the makeup air is energy intensive. Careful attention to exhaust systems, hours of operation of exhaust, and all cooking equipment can yield savings. In addition, commonsense practices in kitchen design, construction, and operations include locating heat-producing equipment far from refrigeration units, turning off equipment when not in use, and purchasing the most efficient equipment possible. Gas-fired equipment is generally more efficient with lower climate-altering gas emissions per Btu than electric equipment. Heat recovery is often difficult from kitchen exhaust due to grease in the air, but where possible, cleaner exhaust streams may provide that opportunity.

Dining facilities and service lines also offer opportunities to save lighting energy, including reducing wattage from overhead and accent lighting as well as shutting equipment off (or not turning on until needed)—particularly at service tables. At Tufts, we have been successful in installing compact fluorescent lighting in walk-in refrigerators and above salad bars and other service spaces. Incorporating energy-efficiency training as part of dining-worker training can be helpful as well.

Tel-Data Spaces

The infrastructure for handling computers and telephones has exploded on college campuses in the last decade. In new campus buildings there are telephone and data (tel-data) rooms that serve an individual building or large areas of campus by housing computer hubs and data equipment—all of it heat generating. In many cases, dedicated air-conditioning is provided to these spaces year round. As with any air-conditioning, there is a need to rightsize this equipment, and careful attention to the actual climate needs of the equipment can help to improve efficiency.

One of the major efficiency issues facing tel-data spaces is preventing "recirculation" of warm air that is mixed with conditioned cool air. Telephone and data equipment is generally arranged in rows. Significant savings can result if air-conditioning systems use a "hot aisle–cold aisle" layout. This layout is designed so that the front of all servers face one aisle (the "cold aisle"), while the exhaust faces the next aisle (the "hot aisle"). Cool air is routed down the "cold aisle" and is pulled through the server blades to be exhausted onto the "hot aisle." By appropriately arranging the data-equipment storage space using this concept, costs can be reduced by approximately 20 percent.[12] In this way, the exhaust from the systems is concentrated in one aisle (computers are arranged back to back) and the cooling is provided on the other. In cooler climates, glycol cooling may provide a significant efficiency benefit for tel-data spaces. With these systems, once the outside air is cool enough, the system provides "free cooling" by bypassing the heat-exchanger unit of the air conditioner and shutting down the refrigeration cycle.[13]

Equipment in Buildings

Electricity using equipment such as computers, cell phone chargers, printers, copy machines, televisions, hair dryers, and room refrigerators creates "plug" loads, or electrical loads. In the last decade this has been a significant and rapidly growing portion of a campus electricity profile and a contributor to a growing need for air-conditioning. In the early 1990s Tufts undertook two major retrofits of lighting in Halligan Hall, our electrical engineering building. We advocated for these projects based

outweigh the gains by modifying practices of building users or by changing equipment in individual buildings.

The central boiler is a logical place to look for efficiency gains and emission reductions. The various operating aspects that deserve attention include:

- Equipment scheduling and operating practices
- Boiler-plant efficiency measurement
- Draft control
- Air-fuel ratio
- Burner and fan systems
- Combustion gas heat transfer and heat recovery
- Condensate, feedwater, and water treatment
- Fuel-oil systems
- Steam and water leakage
- Conduction and radiation losses
- System design for efficient low-load heating[14]
- Fuel switching (to a less carbon-intensive fuel)

As with the boiler plant, the chiller plant is another location where small gains in the central cooling plant can create large benefits. Particular areas that should be noted include the following:

- Equipment scheduling and operating practices
- Optimum operation temperatures
- Condenser and evaporator heat-transfer efficiency
- Heat-rejection equipment
- Pump energy consumption
- Compressors
- Refrigerant condition
- System design for efficient low-load cooling
- Exploiting low ambient temperature for water chilling
- Heat recovery from chillers
- Cooling thermal storage[15]

The system of distributing steam, hot water, or chilled water throughout a campus requires maintenance and care to ensure the losses are reduced to the extent possible. A detailed discussion of these issues is outside the scope of the book. However, the scale of the opportunities to reduce climate-altering gases from central heating and cooling systems, as well as from the on-site generation of power, should not be overlooked and may be more important than any other single effort.

Combined Heat and Power

Combined heat and power (CHP), also known as cogeneration (or cogen) technology, combines electrical and mechanical equipment into an operating system designed to convert fuel energy into both electric power and useful thermal energy.[16] CHP systems can be as small as a few kilowatts or as large as several hundred megawatts. While companies have explored home cogen units, the most likely applications in the near term are for commercial or institutional use. Modern CHP systems are typically based on the latest turbine technology, which has extremely low emissions compared to those used even a few years ago. When CHP replaces central-station power generation, carbon emissions are typically reduced by 30 percent.[17] The significant efficiency gains make CHP a good option for many institutions; however, CHP faces some regulatory and practical considerations that make a case-by-case evaluation critical.

Because many decisions are required when evaluating the feasibility of cogeneration technology at a particular installation, it is important to keep in mind the diverse nature of the technical and economic issues that need to be considered in the decision-making process. Evaluating cogeneration potential is a multistage process that begins with an understanding of the infrastructure and operating requirements of the installation. A strategic element is the identification of facility goals and objectives that can be used as screening criteria throughout the evaluation process. For some installations, the primary goal is to reduce the amount of purchased electricity and replace it with lower-cost electricity generated on-site. For others, reliability and reducing peak loads are goals. Tools and professional resources are available for obtaining site-specific data; identifying energy-saving opportunities; performing preliminary screening assessments; preparing preliminary designs; conducting detailed screening analyses; addressing health, safety, and environmental issues; and understanding electrical grid interconnection issues.

Although implementing a CHP project is not technically feasible at every campus and the payback may be somewhat longer than other steam-system improvement options, many boiler owners and operators find cogeneration technology to be a cost-effective savings opportunity over time, and it may have the additional benefit of providing power when the grid is down.

Numerous universities and colleges throughout the country use CHP systems.[18] These range in size from 150-kilowatt (kW) to 500-megawatt (mW) systems. They also include a range of applications from replacement of existing steam and electric generation to systems that serve campus expansions. Some of these systems are run by the university itself and others are partnerships between the institutions and companies that build and operate the plants for the universities (such as the plant at the University of Maryland at College Park). Because of the high efficiencies and the emission reductions, combined heat and power is an important technology for any college or university to consider seriously. While most cogen plants are invisible to the university community, MIT's cogeneration plant is centrally located and highly visible.

Cogeneration systems are also available to small-scale users of electricity. Small-scale packaged or "modular" systems are being manufactured for commercial and light industrial applications. Modular cogeneration systems are compact and can be manufactured economically. These systems range in size from 20 to 650 kW.

Capacity from Efficiency

Central heating and power systems, both on campus and off, have a finite capacity for delivering service. Metropolitan areas, such as Boston, may have "congestion" areas—areas where there are sometimes problems delivering sufficient electricity to meet demand. Campuses may have congestion problems if several buildings are served by a central boiler and chiller; there may not be enough capacity to serve an addition or nearby new building. If climate goals are to be met, reducing existing demand and freeing up capacity is a strategy that should be considered and may be cost effective. To identify these opportunities we need to expand the boundaries of construction projects, and of the consulting services that inform them, beyond the project's walls so they include an analysis of electric and heating capacity from efficiency in existing demand rather than just new sources.

Tools for Green Buildings and Climate Change Action

Numerous resources are available for members of the university community interested in green buildings, energy efficiency, and climate change action. This section describes some of these resources.

Professional Community

The most important tool for green buildings and climate change action is the professional community and professional literature. Advocates working in this area should not underestimate the vast expertise of experts and the comprehensive literature on energy efficiency and green buildings. Conferences, listserves, training sessions, and trade magazines are important sources of current information.

The American Society of Heating and Refrigeration Engineers (ASHRAE) develops many industry standards and best practices. The American Physical Plant Association (APPA) is an association of college and universities physical plant managers and directors. The Society for College and University Planners, the National Association of College and University Business Officers, and other college and university professional organizations are devoting significant resources to these issues. Outside of the higher education professional associations there are countless organizations and government agencies dedicated to providing energy efficiency advice and technical help. Engineering consultants and other building professionals are useful and can be particularly helpful in developing applied strategies for your campus buildings. Staff engineers and facilities personnel are also invaluable in finding solutions that can work for your college or university.

The literature from the environmental community is often more general. It is particularly helpful in demonstrating that other institutions are embracing issues of sustainability, green buildings, or green power, but is not a substitute for applied knowledge specific to the particular issue. Student groups often gain good ideas from groups on other campuses from these sources, but climate action advocates should be careful not to overlook the complexity of most build-related measures.

Utilities staff remain an important part of the professional energy community. In many areas utilities still provide incentives for efficiency and expertise (as well as financial help for obtaining expertise) on efficiency measures. Other government agencies at the state and regional level also have vast technical resources. The DOE EERE State Activities & Partnerships website provides useful links to partnerships with and projects in the states.[19]

Leadership in Energy and Environmental Design (LEED)

The U.S. Green Building Council (USGBC), a coalition of leaders from across the building industry, created Leadership in Energy and Environmental Design (LEED), a certification process for green buildings. LEED was created to define *green* by providing a standard for measuring building performance. Its goal is to set a design guideline "to promote full building design, and to establish a market value for the green standard." LEED provides a tool (the Checklist) and a certification process to provide some guidance and structure to "green" buildings by establishing performance goals and industry standards. It has built-in flexibility and recognizes the trade-offs inherent in building projects.

The LEED process assigns points in six categories: Sustainable Sites; Materials & Resources; Water Efficiency; Indoor Environmental Quality; Energy & Atmosphere; and Innovation & Design Process. While climate change advocates will focus mostly on the energy and atmosphere requirements for a building, there are many opportunities for responsible design that are beyond the energy focus, some of which have energy implications. For example, the strategies in the sustainable-sites category include alternative transportation, light pollution, site selection, and landscaping. Each of these has consequences for reducing climate change impacts. In the water-efficiency category, saving water as well as innovative wastewater technologies can themselves save energy. The construction and waste management targets can reduce climate change impacts from solid waste disposal. The materials-and-resources category can increase use of recycled materials and thus reduce the embodied energy in a building. Selecting local resources, those from within a 500-mile radius, can also reduce transportation impacts. Other categories such as indoor air quality require low-emitting materials like paints and carpet. These are good construction practices that will benefit the building occupants for years to come. Day lighting and views also have energy implications.

Although not perfect, LEED has become an accepted industry standard. Architecture, engineering, and trade associations routinely highlight green design elements and LEED as a tool in their literature, conferences, and professional development. Across the country, college and university buildings are using the LEED standard to improve their design. Some universities, such as Emory University, MIT, and the University of California system, are using LEED as a standard for all buildings. Emory University's board of trustees recently approved the LEED standard for construction of all projects at the school. Emory's Whitehead Biomedical Research Building was the first building in the southeastern United States to earn a LEED rating in 2002. The university plans to submit at least ten buildings, or a total of about 1.1 million square feet, for LEED certification.[20] The University of California system has drafted a policy that all buildings (except acute-care facilities) shall be built to the LEED 2.1 Certified status, with a recommendation that the buildings be built to at least LEED Silver status if financial resources allow.[21]

The USGBC has also instituted a LEED-EB (Existing Buildings) program. This program applies to buildings that are not undergoing a major renovation, which would qualify for LEED-NC (New Construction). LEED-EB is designed to increase operational efficiency while decreasing environmental impacts associated with building management. The program addresses energy and water use, indoor air quality, recycling programs, and systems upgrades.[22] Improving the efficiency and maintenance of existing structures can provide a significant reduction in greenhouse gas emissions; this program provides building managers with a system for doing so.

LEED is a tool that encourages comprehensive green building design and provides a framework for this discussion. However, LEED does not address issues of sufficiency or the fundamental need for building, which is a critical issue if we are really to address emission reductions comprehensively. LEED also addresses only energy use per square foot, but does not address energy use per student or per research dollar—metrics that may be more suitable for the university environment. Lastly, LEED buildings may have only modest energy-efficiency measures and still qualify. Nonetheless, this industry standard is a good tool for helping a design team to set design goals and to evaluate trade-offs throughout design and construction.

Energy Star

Energy Star is a government-backed program, administered by the Environmental Protection Agency and the Department of Energy. The voluntary program raises awareness among businesses and individuals of energy-efficiency opportunities in buildings and appliances. Energy Star programs provide training, product advice, comparative building analysis, energy-efficiency tools, and guidance materials.

Energy Star building rating systems focus on energy consumption, but require an indoor air quality assessment as part of the certification process. Colleges and universities can participate in this program and receive certification that their building is compliant. The process of seeking and achieving an Energy Star rating for a new building and/or renovation can help to improve the design and construction process as well as call attention to the institution's efforts. Compared to the national building stock, Energy Star offices are approximately 40 percent less energy and cost intensive than average buildings, according to EPA. These buildings have achieved energy savings while maintaining indoor environments that have been professionally verified as compliant with current industry standards.[23]

Design Standards

Achieving green design—whether it is achieving LEED certification or achieving effective energy conservation—requires careful attention to incorporating the design into the specifications for the building. Specifications are the documents that tell a contractor exactly how to build the building. Design standards are university statements of purpose and intent and/or specific prescriptive requirements for all relevant building components in order to help standardize systems and ensure high quality. Standards can be instrumental in ensuring energy efficiency as well.

Most design specifications use the Construction Specification Institute (CSI) format as a guide. This format divides the specifications into divisions. Adding green design and energy efficiency throughout the design guidelines and resulting building specifications will help to ensure that the desired energy efficiency is achieved. However, the most obvious opportunities for climate change action are in the general sections of the specifications and the mechanical, electrical, and building envelope sections.

Design standards are essential for any successful building project since they lay out the design expectations and specify campus standards for all building components from toilets to energy management systems. A great deal of time can be saved if standards specify equipment and design intent. Many institutions have incorporated green building standards into their existing standards. Some have overlaid a separate green standard on the existing standard. Many design contracts reference the design standards, so the standards are a binding part of the working relationship. Regardless of the approach, incorporating green elements, with a focus on energy efficiency, into the standards is critical. The Federal Energy Management Program has extensive free standards that are a good jumping-off place for any institution.

Energy Simulation

Creating a mathematical model of how a building will use energy can be a very effective tool for making decisions. However, in order to provide maximum benefit, the model, even in its simplest form, needs to be up and running during the schematic design phase of a building project. Models are decision-making tools that allow you to see the effects and interconnections among various building components and to compare relative merits of choices. While they have a margin of error in predicting actual energy use, models are a good tool for understanding tradeoffs and the implications of certain design decisions. They should be done early in the process to examine scenarios and inform design. Later in the process models will be more specific, but design decisions have less impact on energy as the design becomes finalized.

Understanding the assumptions being made is a critical aspect of developing a good model. For example, assumptions about building use and occupancy influence how many hours each day lighting and HVAC systems are on. A good model will be accompanied by a narrative that lays out the assumptions that have gone into the model. And models must be calibrated with real-life results. Models need to be checked by hand to quickly determine if the Btu per square foot per year and the square foot per ton of cooling are in the right range. Most importantly, energy models should be used to test design alternatives and provide feedback that informs design. They must be a complement to common sense and experience, not a substitute for it.

Project Oversight

Throughout design and construction, the university must exercise detailed oversight to ensure that systems meet the institution's needs. Catching mistakes, overdesign, or underdesign early, before they are buried in construction documents, will make problems easier and less costly to fix. A qualified and experienced project manager can probably save his or her salary many times over in the course of a project by keeping a careful eye on the process. Climate change advocates can help take responsibility for overseeing and ensuring quality in the areas they are interested in whenever possible.

Commissioning and Recommissioning

Commissioning is a process of auditing a building or renovation during the design process and the construction phase to ensure that it will work and is built as designed. This is particularly important for energy-intensive elements such as HVAC systems. A commissioning agent works for the university to evaluate the design documents to find and fix errors, identify more efficient and effective ways to solve problems, and catch problems before they become crises. Commissioning agents then observe and verify construction to ensure the same result. To be done right, commissioning agents should be hired early in the design process to examine all elements of the design. It is, however, more typical for the commissioning agent to enter the process well into final design or even in construction. Regardless, the commissioning agent is a valuable resource advocating on the owner's behalf.

Recommissioning is a comprehensive audit of an existing building to determine if the building systems are operating as designed, if they are operating optimally, or if the building systems need to be updated or changed to keep up with building-use changes. Tufts has found the recommissioning process helpful in identifying important energy-saving opportunities in a variety of building types from libraries to research laboratories. These opportunities include the use of new technology, fixing systemic problems, dealing with new building uses, and adapting to new space-use patterns.

Benchmarking

Building benchmarks provide comparative analysis of buildings by type and use. Benchmarking generally compares building energy use on a per square foot basis for particular building types. Energy Star provides some benchmarking tools for comparing your buildings against those of similar building types.[24] Other sources of benchmarking include intra-campus building comparisons of like buildings or regional campus-to-campus studies. These analyses can validate engineering claims that buildings are or will be efficient and can be useful for identifying problems for further study.

Maintenance

Building maintenance staff are critical to efficient ongoing operations of building systems. Routine maintenance such as cleaning filters, lubrication, and regular inspections can help to prevent problems and ensure that systems operate as efficiently as possible. Cleaning of lighting and refrigeration coils can improve efficiency as well. In studies of Energy Star buildings, there is ample evidence that even the presence of highly efficient equipment alone is not the sole predictor of building performance, implying that good building management and maintenance can also significantly contribute to building performance. Maintenance is essential for ensuring that a building works as it is designed.[25]

Trained maintenance personnel can also be the eyes and ears of an energy manager—finding and reporting problems. Empowering maintenance staff and providing incentives for their good ideas is essential. Ongoing training is also essential for maintenance personnel so that they can stay abreast of new techniques, technologies, and methods in the buildings that they service.

In new buildings, a maintenance plan and a building turnover plan should be comprehensive. It should provide hands-on training for staff, a video of the training session so new employees can be trained easily, systems manuals, and maintenance schedules.

Building Curators

Developing and implementing a building curator program is one way to increase the number of trained eyes in existing buildings. Building curators are designated employees in each building that are trained to identify building-related problems and report them to their university facilities departments. The most successful programs can help to report and solve problems quickly, communicate effectively with employees, and troubleshoot problems with facilities maintenance crews.

Curators are responsible for learning how the building systems function, performing simple systems checks and adjustments, collecting data in some cases, and identifying problems and communicating them to the maintenance staff. Establishing an individual responsible for each building will help occupants understand that the institution is committed to building issues, though not all problems can be solved at once. For a program of this type to be effective, the maintenance staff has to be active and enthusiastic supporters. This support may be relatively easy to gain in cases where building curators can minimize the number of calls to report problems in a building and can report with sufficient accuracy that maintenance staff arrive with the proper equipment and replacement parts.

The Role of Buildings That Teach

University buildings are traditionally considered support systems—places to teach, learn, eat, sleep, or exercise. We rarely notice the building until it is uncomfortable, cramped, or otherwise failing to support its use. But buildings offer unparalleled opportunities to teach, and few of our buildings are maximizing this learning potential. Among the disciplines that can find rich teaching material in a building are:

- Business, finance, and economics
- Structural, civil, mechanical, and electrical engineering
- Thermodynamics
- Hydrodynamics
- Physics
- Air, water, and soils pollution and treatment
- Agriculture, nutrition, and health

- Physical accessibility
- Policy, politics, and community relations
- Foreign languages (often workers are nonnative speakers of English)
- Project management, security, and fire protection
- Traffic and noise engineering
- Energy technology
- Climate change and air quality
- Many more

Universities miss most of these learning opportunities that exist right in their own buildings, often only touching on them in "green" buildings of the recent past.

A Civil and Environmental Engineering (CEE) course at Tufts called "Engineering and the Construction Process" was cotaught by a CEE faculty member, Chris Swan, and the project manager from the construction firm managing new construction on our Medford campus. The course met on-site and used two construction projects to teach students about the process and the engineer's role.

Summary

Buildings and university facilities must be a focus area for any comprehensive effort to reduce greenhouse gas emissions. At most institutions, action to improve energy efficiency and operate buildings is underway at some level since the potential to save money can be significant. Nonetheless, many opportunities for efficiency, fuel switching, and alternatives in new and existing buildings probably remain.

Considering buildings as an integrated system rather than a collection of unrelated parts is a critical shift. In many design and renovation projects the largest savings may be realized by finding opportunities for optimizing systems, often by identifying their interrelated components.

8

Action beyond Facilities

Even if all of the buildings on campus were very efficient, there would still be many opportunities to reduce emissions through purchasing decisions, transportation options, and land management and other aspects of university operations. In this chapter we examine many of these opportunities to address both primary and secondary emissions. Some of the actions described in this chapter are highly visible but largely symbolic. Others have the potential to reduce emissions significantly and create market incentives for larger change as well as to create awareness. To the extent that visible actions are used as education activities, they may have multiple benefits. Still other efforts, such as a switch to biodiesel fuel, may have a minor climate change benefit, but will have other benefits such as reducing reliance on petroleum-based products. Some of the actions apply primarily to institutions with a specialty in agriculture, where there is a large herd present, or forestry, with access to university-managed forests.

At Tufts we initially focused on our buildings and our energy sources as the first tier of action. As our effort matures, we have expanded to other realms of institutional operations that have emission reduction potential. While we do not present a complete action plan for any of the issues listed, we intend to help develop a broader understanding of the extent to which climate change is affected by a wide range of decisions on campus.

University Purchasing

Universities are consumers, and like all consumers, their purchasing decisions can underscore their priorities. University purchasing decisions offer opportunities to have vendors, contractors, and service providers improve their practices on the university's behalf. Many campus sustainability programs have benefited from projects that involve the purchasing department and the university's purchasing power in efforts to increase recycling, increase the purchase of recycled content of materials, reduce waste, increase use of local materials, and reduce toxics. Programs such as those at Rutgers[1] and the Massachusetts Operational Services Division[2] have made extensive and effective use of requests for bids, product and service specifications, and contracts to meet their program objectives. These same tools can be used to reduce emissions of climate-altering gases generated as a result of the university's purchasing. These "upstream" or secondary impacts result from energy input to materials, transportation, manufacturing processes, and the specification and selection of products themselves.

Upstream CO_2 emissions are enormous. In a recent investigation of the nearly 500 sectors of the U.S. economy, the following surprising results were found:

- For a majority of sectors, upstream emissions exceed direct emissions.
- For many sectors, upstream emissions are five to ten times direct emissions.
- The largest sector in terms of upstream emissions is the construction sector, with upstream emissions five times its direct emissions.[3]

Upstream greenhouse gas emission releases are not as readily apparent as those from heating and electricity production, nor are the actions to address them as obvious or within direct control of decision makers. Production activities elsewhere in the economy and their respective environmental burdens are linked to every university by purchasing decisions. And likewise, secondary purchases or higher tiers of activities are linked by purchases from primary commodity providers. The links continue to tertiary and higher-order tiers, or links may exist to reuse or recycling loops in the economic system. These activities represent the economic system in totality. They include extraction of raw materials, production of intermediate and final commodities, and all of the services (e.g., legal and technical consulting) required to produce such commodities. Educating the student body at Tufts requires a wide range of direct and indirect purchases: construction of buildings, furniture purchases, photocopies, boxes of markers, reams of paper, and even landscaping. Each purchase of a commodity is embodied with the upstream production activities required to produce it—creating the link to the associated emissions. This is the essence of input/output economics in a life-cycle context—the ability to capture the full spectrum of production activities associated with any commodity purchased without arbitrarily truncating the upstream boundary due to real-world costs and inaccessibility of data.[4]

A university can buy green power, recycled products, fuel-efficient vehicles, and local food—all which can have lower associated emissions. The purchasing power of a single university may affect decisions or even the viability of a small provider, and the combined purchasing of several universities (or other institutions such as schools or local governments) in a region can influence markets. Comprehensive climate change action can mobilize university purchasing power to address some of these upstream impacts, essentially internalizing some external costs. Put more bluntly, colleges can put their money where their mouth is.

Purchasing Electricity

There are two major parts to the delivery of electrical power generated outside of a university: generation (creating the electrons) and transmission (bringing them to the institution). At Tufts, as at many institutions, our electricity is produced off-site. Electricity is deregulated in Massachusetts, which gives Tufts, and all other electric customers in the state, the opportunity to purchase the generation of electricity from any producer. However, the responsibility for transmission remains with our traditional utility companies (Tufts has two). At present, the market in Massachusetts offers few generation choices, and like most institutions Tufts selects generators based largely on price, although recently emissions were introduced as a consideration.

Conventional electricity generation causes a range of impacts to the environment. Fossil-fuel-fired power plants are responsible for 67 percent of the nation's sulfur dioxide emissions, 23 percent of nitrogen oxide emissions, and 40 percent of human-made carbon dioxide emissions.[5] These power plants are also the main source of mercury emissions in the United States.[6] In addition to the direct emissions associated with fuel combustion, there are also emissions and other environmental effects associated with the extraction of oil and coal and the disposal of fly ash from coal combustion. Nuclear power plants produce electricity with negligible direct carbon emissions (although they have some emissions in the fuel cycle); however, the siting, safe operation, and sound disposal of waste from these plants remain a concern for many in existing or prospective host communities. Both nuclear and fossil-fuel plants that serve the grid are large physical facilities with considerable embodied material. Depending on the fossil fuel used, there will be a range of impacts and risks associated with transporting the fuel to the power plant, whether this is accomplished by pipeline, tanker, truck, or train.

When the generating or collecting equipment for renewable energy is manufactured, there are waste products. Wastes generated in the process of manufacturing solar arrays are modest in scale and roughly comparable to waste generated by the semiconductor industry. Similarly, wastes are generated in manufacturing wind and hydropower turbines; however, these are also modest in volume and are roughly comparable to wastes generated in the aircraft industry. In both the semiconductor and aircraft industries, considerable efforts have been made to develop processes to minimize wastes. The generation phase of renewable energy offers an alternative with either zero or significantly lower emissions than conventional power sources.

The advent of a deregulated electrical industry makes it possible for customers to select their generators based on a variety of factors, including the amount of pollution or heat-trapping gas emissions from the fuel that generates the electricity. (Deregulation is a state decision and does not exist in every state at this time.) Electricity generated from hydropower or nuclear power has fewer greenhouse gas emissions than that generated by a coal-fired plant, and that could be a decisive factor. Nonetheless, a consumer could decide that the destruction of fish spawning grounds from hydropower or the generation of radioactive waste from nuclear power was unacceptable and select power from wind, solar, or landfill methane recovery. Consumers may also be able to choose electricity generated from natural gas instead of coal, or from wind or solar. Presently, however, these choices typically come at a premium price, although longer-term contracts are providing lower costs paired with lower emissions in some circumstances. In the near term, the greatest opportunities for reducing emissions are likely to involve utilizing natural gas, the fossil fuel with the lowest carbon content per unit of energy. At this writing, however, natural gas prices are high and in New England, supplies may be insufficient to meet the increased reliance on natural gas for the growing winter demand.

Some states have implemented a renewable portfolio standard (RPS) for their local utilities. This requires utilities operating in the state to generate a specified portion of their electricity from renewable sources. The amounts, target dates, and allowed fuels vary by state. These RPS provisions have been instrumental in helping to increase the use of renewables and develop credible methods for tracking them.

Purchasing Green Power *Green power* is a marketing term for electricity that is partially or entirely generated from environmentally preferable renewable-energy sources with low or zero emissions of climate-altering gases and includes solar, wind, geothermal, biomass, landfill methane, and low-impact hydro. Green power from landfill methane has an added benefit of reducing emissions of methane, a more potent climate-altering gas, into the atmosphere. Green power may be offered as a delivered product (the electrons into the grid) or as "green certificates." Renewable-energy supplies are growing around the world, with wind energy now meeting up to 20 percent of the energy needs of Denmark and some of the needs of Germany and Spain.[7] Electricity suppliers offer electricity from renewable resources either as a percentage of electricity use or in a fixed number of units or blocks (usually 100 kilowatt-hours). Either way, most green power customers end up with a blend of renewable and conventional power. About 50 percent of retail customers in the United States now have the option to buy green power directly from electricity suppliers serving their area.[8]

The burgeoning market for renewables has created a way to separate the green attributes—the good things about green energy, including reduced CO_2 emissions—from the electricity itself. This market makes it possible to pay extra for the "greenness" or "green attributes" of the electricity as distinct from the electricity itself. Renewable energy certificates (RECs)—also known as green tags, green energy certificates, or tradable renewable certificates—represent the technology and environmental attributes of 1 megawatt-hour of electricity generated from renewable sources. Figure 8.1 illustrates the way that green certificates

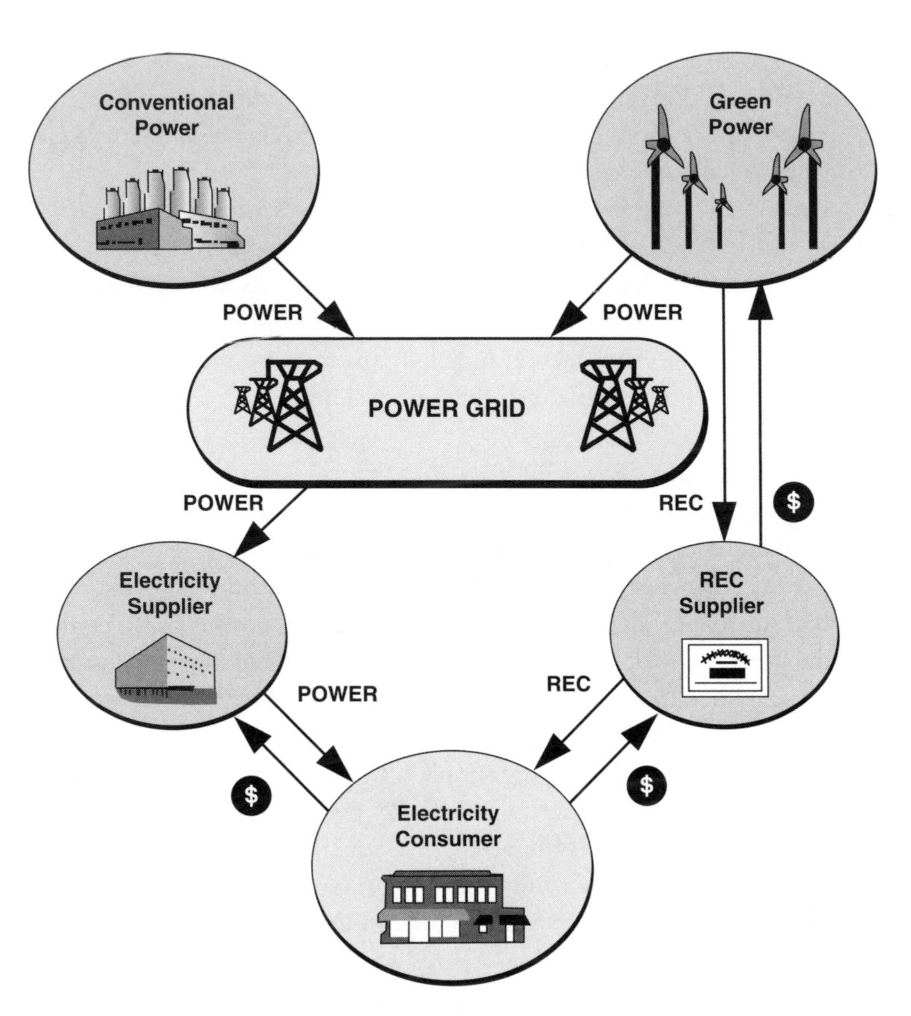

Figure 8.1
Elements of purchased power *Source:* U.S. Environmental Protection Agency, *Green Power Purchasing Guide*, http://www.epa.gov/greenpower/pdf/purchasing_guide_for_web.pdf

or renewable energy certificates (RECs) work. These attributes may be sold separately from the associated electricity. If the attributes are separated from the associated electricity, the electricity is no longer considered "green."

Because RECs are sold separately from electricity, they can be purchased from locations anywhere, enabling organizations to support green power even if their local utility or power marketer does not offer a green power product. Customers do not need to switch from their current electricity supplier to purchase certificates, and they can buy RECs based on any fixed amount of electricity. Certificates can be purchased from REC marketers or, in some cases, directly from renewable-energy generators. Because certificates can come from energy generated in any geographic area, the location where electricity was generated—and therefore the area in which the environmental benefits are likely to accrue—can be an important factor to consider.

The delivered green product is currently less common, but electricity products made with green power always contain a higher percentage of electricity from renewable-energy sources than conventional electrical service does. These products are often a blend of generation sources in order to make them cost competitive. Certification efforts such as those provided by Green-e (see box 8.1) are available to ensure that the product can be properly described as "green" and that the electricity that is generated is as advertised. Purchasing this product ensures that a

Box 8.1
Green-e

The Green-e Renewable Electricity Certification Program is a voluntary certification program for renewable-electricity products sold in competitive electricity markets. The Green-e logo helps consumers identify certified renewable-energy products. (For further information, go to www.green-e.org.)

defined amount of green electricity generated from the fuel mix is "delivered." However, it is important to clarify that this electricity is delivered into the regional utility grid, rather than to any particular customer. Nonetheless, purchasing this product helps to create a market for green power and ensure its presence in the market.

When an individual or an institution elects to switch electric generators, they experience the conventional level of reliability in power delivery since the electricity is delivered into the grid, rather than to the user directly. In addition, the local utility will remain responsible for delivering power (transmission) and maintaining the wires. In the event of a power outage, the local utility remains responsible for problem solving.

At present, green power almost always costs more than conventional power. For that reason it is important to consider the reasons for purchasing the power and to make sure that the product purchased has the attributes that the institution seeks. Some important considerations that can affect price include:[9]

Location Local generation of green power can help improve local environmental conditions and foster local markets for renewables. However, some resources, such as wind, are not distributed evenly across the country, making wind investment prices vary with their location. Products such as landfill methane gas facilities or wind turbines can be linked to a specific generation site, and thus can be identifiable and tangible for the campus community.

Product mix Products vary with respect to the amount of renewable power and the types of resources in the product. Prices may reflect the amount of renewables in the product.

Additionality New or incremental renewable sources are facilities that are additions to the market, rather than the existing facilities. Presumably, existing facilities sold power into the grid before a particular university purchased the green power product that includes them. However, in some cases, support for existing facilities can ensure their continued operation in the face of market pressures.

Length of contract There may be market advantages to longer contracts. In addition, longer contracts may provide opportunities to lock in long-term prices that allow utility managers to predict costs with increased accuracy. However, the electricity market is changing rapidly, so universities will want to maintain some flexibility and to read carefully all of the fine print.

Certification It is important that an investment in green power have a verifiable means of ensuring that the power is there. Green-e is one certification process.

How Much to Buy? Ideally, the university will decide to buy green power for all of its electricity needs, or to buy RECs to ensure green attributes for all its power. The College of the Atlantic in Bar Harbor, Maine, has opted to do this.[10] However, this is often cost prohibitive, and institutions are more likely to buy just a portion of their power in this way. Many institutions link the purchase to a specific building or activity or event.

Recent experience at Tufts indicates that changing to a longer contract period may allow for cost-neutral or even less expensive electricity generation from renewable sources. In our case, the existing hydropower utility knows its costs well into the future and can enter into contracts over long time horizons.

Paying for Green Power Because green power usually has a premium price, decisions must be made on how to pay for this "luxury product." Across the country universities are using a combination of funding mechanisms. These include:

Student fees At Western Washington University, students voted to increase fees and are using the revenue to purchase RECs. This tactic may prove successful when the number of RECs is tied to a particular goal or a specific building or school. At Western Washington, the goal was 100 percent renewable sources. A high percentage (84.7 percent) of the students voted in favor of the fee increase, and as a result, the university buys about 35 million kWh from renewable sources.[11]

Administrative funds The University of Pennsylvania's administration is purchasing RECs to offset 10 percent of the university's electricity use.[12]

Combination Some schools are using a combination of student fees and administrative support.

Savings from energy projects Energy-efficiency projects can have substantial savings, both from reduced operating costs and from rebates. These savings can be pooled and reinvested in other energy-related projects or in green power.

Donations Donors may be very attracted to the idea of making it possible for the college to increase its environmental profile by purchasing green electricity.

Creating a Hedge on Utility Prices with Renewables Some renewable sources, such as solar and wind, have fixed and predictable costs. Since the current and future price of wind (the fuel) is known and zero, the costs of generating wind power are limited to the initial development and maintenance costs, and are therefore easier to predict than the price of a barrel of oil or the price of natural gas. Developing a long-term relationship with a wind producer may allow a university to negotiate an advantageous and fixed price for a ten- to fifteen-year time period. While the cost of wind power may be higher than that of fossil fuel in the early years, it may be lower than for conventional sources in the long run as prices of coal, oil, and natural gas rise.

Contracts for Goods and Services

Contracts for goods and services are ideal opportunities to specify strategies or products that are energy efficient or otherwise reduce climate-altering gas emissions. Contracts usually offer centrally bid or negotiated terms for the good or service provided, and energy efficiency should be a key component of the specifications. Some opportunities for contracting include:

Construction and facilities contracts Contracts for architectural design and engineering as well as for construction should clearly state expectations around energy use, efficiency, and technology (see chapters 6 and 7).

Maintenance and service contracts Equipment and facilities maintenance contracts can provide opportunities for personnel to identify energy savings as well as expectations that equipment will be replaced with the most efficient models.

Dining-services contracts Third-party companies run many dining operations, and including energy performance and incentives for high performance in their contracts can yield savings.

Washing-machine contracts Front-loading washing machines use less energy and water than conventional machines. At Tufts, switching to front-loaders is anticipated to save $30,000 and reduce carbon dioxide emissions by 30 tons annually.[13] While the switch may not always be logistically easy (some front-loaders are wider), it is well worth investigating in the contract.

Vending contracts Vending machines use electricity continuously. At Tufts we found that a typical soda vending machine uses about $381 of electricity a year (more at current prices). To address this, we began installing VendingMisers®, devices that cycle the vending machine's compressor on and off less frequently. The payback on these devices was about one year;[14] however, we had problems in communicating our expectations to the vending contractor and the misers were frequently disabled, reducing their effectiveness. New Energy Star vending machines are available on the market, and while they may not save as much energy as a properly operating vending machine equipped with a miser, they are likely to have a higher success rate when many machines are scattered throughout the campus. Vending contracts can specify them.

Transit-service providers Many colleges and universities contract for shuttle bus and transit services. When this is the case, there are opportunities to negotiate for alternative fuels, strategies to increase riders and decrease single-vehicle use, and ways of using vehicles that are sized to accommodate demand on each route.

Copier contracts The energy use of copy machines varies considerably. University contracts should stipulate that copy machines be Energy Star–rated and have a sleep feature.[15] Copy machines should also accommodate recycled paper and have duplexing features (making back-to-back copies) to reduce paper use. Copier mechanics should be required to set energy-saving features to the shortest settings as well.

Computer contracts As we have discussed, the growth of computer equipment on college campuses is enormous. Stipulating Energy Star computer, sleep-mode features, and flat-screen technology can help to reduce the energy use from this equipment.

Purchasing Energy-Efficient Products

Whenever possible, energy efficiency should be a criterion for purchase. While it is important to conduct life-cycle analyses to determine if any price premium is worthwhile, the results of these analyses are highly dependent on the assumed energy prices over time—price increases in late 2005 and early 2006 changed the results of these analyses very quickly.

Energy Star, a government program to promote energy efficiency, offers ratings for a variety of appliances and office equipment based on the energy consumption of the equipment of similar size and type.[16] This rating should be a minimum standard for university purchasing. In addition to acquiring Energy Star–compliant products, university purchasing should look to downsize where possible. Smaller is generally more efficient.

Factors beyond Cost

TCI undertook several analyses of energy savings for Energy Star–rated appliances. Our goal was to evaluate whether the increased first costs of several different technologies of computer monitors, computers, and copy machines were offset by electricity savings during the operational phase. In our first effort we looked at making flat-screen computer monitors the default option when purchasing a computer package from the university's supplier. At that time, flat screens cost more than conventional CRTs but used less energy. We thought that perhaps the more expensive screens could be justified if full life-cycle costs were taken into account. This turned out not to be the case given the assumptions we used about electricity prices and hours of operation. But to our amazement and delight, many flat screens are appearing on campus.

The flat screen is a good example of a product for which people are willing to pay a premium (although prices are quickly dropping). We point this out to debunk the myth that all decisions on campus need to be justified on the basis of their first cost or even their life-cycle cost. People like features of the technology, and flat-screen monitors take up less space and have an updated look. Because so many other climate change actions are scrutinized closely, and must be sold and defended on a first-cost basis, it is important to note that there are countless other times when first cost is not the only consideration in decision making and may take a backseat to aesthetics or other attributes.

University Transportation

The transportation sector accounts for 27 percent of total U.S. greenhouse gas emissions and is the sector with the greatest annual growth in terms of GHG emissions.[17] On a campus, transportation-related emissions are produced by university-owned vehicles, contracted transit services, commuters, resident travel, deliveries, grounds equipment, and shuttle buses. The contribution of transportation to a college or university's emissions inventory may be a great deal lower than the national share, and may present a series of challenges for those interested in reducing the institution's emissions. The size and nature of the institution and its location may have a profound influence on the amount of emissions associated with transportation. For example, a commuter college will have a larger share of transit-related emissions. Likewise, many factors dictate the best strategies for reducing these emissions. At Tufts, transportation is an area in which we have done considerable thinking and planning but, to date, have taken comparatively modest actions. Nonetheless, our experience suggests that transportation-related issues are a powerful way to provide education and visibility to a wide audience while reducing emissions.

As with any climate action program, university transportation programs are complicated and their success depends on careful attention to details. The smallest failures can taint the reputation of a program and decrease its use. Will Toor and Spenser Havlick's book *Transportation and Sustainable Campus Communities: Issues, Examples, Solutions* (2004) is a comprehensive how-to guide for developing successful policies. We will not seek to replicate their work, but simply to provide an overview of the opportunities.

Transportation to, from, and around Campus

Reducing single-occupancy travel is a goal of most transportation plans. The strategies are generally a combination of carpooling, public transit, efforts to promote walking and biking, telecommuting, university transit, and parking policy. Increasingly, the process of matching riders with rides is Internet based, and a number of companies will offer that service at very low cost. At Tufts, we have also added a shared-vehicle program, described later in this chapter.

Toor and Havlick suggest that successful strategies to reduce demand for single-occupancy vehicles are a combination of information, facilities, support, and incentives. Table 8.1 summarizes their work and shows examples of each of these strategies for successful programs to promote walking, biking, transit use, and carpooling. Their book describes each strategy in detail. One of the challenges of creating successful programs to reduce emissions from commuters is the complexity of the solutions. Because the programs come to be counted on by members of the community, they must be reliable and consistent. As evidence, Toor and Havlick advocate for a guaranteed-ride-home program—a paid ride home for emergency situations—as part of all of their programs. Their work demonstrates the extent to which successful programs must be fully developed.

Table 8.1
Strategies for reducing single-occupancy vehicle trips to and from campus

Strategy	Information	Facilities	Support	Incentives
Walking-oriented	Guidebook Promotion Walk to school/ work day Maps/routes Orientation Special events	Pedestrian corridors Wide sidewalks Signage Signals Shower/locker facilities Clustered parking Overpasses Pedestrian/bike Campus core	Support group Safety seminars Special events Guaranteed ride home Transportation coordinators Flexible work hours Campaigns Commuter store	Discounts from retailers Commuter club Transportation allowances Parking cash-out Taxation incentives
Bicycling-oriented	Guidebook Promotion Bike to school/ work day Maps/routes Orientation Special events Online bike routes	Bicycle racks Bicycle racks at shelters Bike station Bike lockers Signage Bike paths and lanes Shower/lockers Shared-bike program	Bicycle user groups Bike safety programs Special events Guaranteed ride home Transit coordinators Promotion and campaigns Commuter store	Bicycle accessories Commuter club Transportation allowances Taxation incentives Bike loan program Parking case-out or increased parking rates

Carpool-oriented	Guidebook Promotion Orientation Special events	Vanpool or carpool Preferential parking Site plans include carpool parking High-occupancy vehicle priority system Car sharing	Online rider matching Zip-code meetings Guaranteed ride home Registration surveys Special events coordinators Flexible work hours	Pretax payments University pays cost of empty seat Vanpools First-time ride incentives Prizes and promotions Vanpool subsidy Commuter club Preferential parking and rates for carpools Transportation allowances Van and car loan program For-profit vanpools Parking cash-out or increased parking rates
Transit-oriented	Guide Marketing Maps/routes Orientation Events Promotion Transit routing Realtime transit information	Bikes-on-buses promotion Incidental-use parking Transit-friendly site designs Bus/HOV priority system Campus transit center	Transit to work/school day Flexible work hours Events Guaranteed ride home Coordinators Promotion and campaigns Visitor trip management On-site amenities Commuter store	Taxation incentives Discounted transit passes Pretax benefit for faculty/staff Student transit passes Transportation allowance Parking cash-out Unlimited-access parking permit Rate increase

Source: Will Toor and Spenser Havlick, *Transportation and Sustainable Campus Communities: Issues, Examples, Solutions* (Washington, DC: Island Press, 2004).

Developing this kind of comprehensive program may be easier at larger institutions, where there is established parking and transit infrastructure, than at smaller institutions, where dedicated staff resources do not exist. Likewise, it is easier to implement demand reduction strategies on campuses with limited parking than on campuses with ample parking.

Successful demand management programs require coupling the strategy with parking revenue and parking policy. Identifying the true costs of providing parking with the strategies to reduce the demand is one aspect. These costs include constructing and maintaining parking spaces as well as the opportunity costs for the use of land for parking rather than as playing fields, building sites, or open space. Studies at UCLA showed that the net cost per student of new parking exceeded the cost of subsidized vanpools and subsidized transit-pass programs.[18] Toor and Havlick point out numerous other examples where the results are similar.

It is also critical to manage highly desirable parking spaces to create visible and tangible incentives for climate action. To bring all these pieces together, the university parking officers, police, safety, and human resources staff will have to work together to agree on consistent policy goals.

Student fees can be used to pay for incremental costs of transit programs that reduce greenhouse gas emissions. For example, Ohio State includes a very modest fee for a transit pass as part of each student's fees. This gives the transit authority a revenue stream it can count on, and has increased ridership because students perceive the service as free (they experience no out-of-pocket costs) and because of the ease of use of the program (students simply show their valid university ID in order to ride the system).[19] Ohio State is also interesting because it formed a partnership with the local transit authority to increase bus access.

North Carolina State University operates its own bus system (one of the largest in the state) and also partners with the municipal systems that gives riders access to two other transit systems, CAT (Raleigh's bus service) and TTA (the Triangle Transit Authority, which gives students access to Chapel Hill and Durham, where Duke University is located). In addition, a new light-rail service will have a stop right in the middle of campus, and student fees potentially may pay for a train pass.[20]

On many campuses there are efforts, often organized by students, to consolidate rides home on weekends or vacations. Increasingly students are using the Internet to facilitate these arrangements, developed out of convenience rather than for purposes of emission reduction. Elsewhere campus planning is addressing pedestrian access, bike access, and parking in ways that promote reduced vehicle use.

Shared Vehicles: Zipcar

In summer 2003, Tufts entered into an agreement with Zipcar (www.zipcar.com/tufts). The Zipcar system allows members to rent vehicles by the hour, and is established in urban areas to serve a population that needs an automobile periodically. Nearly 400 members, including administrators, development officers, faculty, staff, and students, can rent vehicles by the hour for university or personal business. While it is envisioned that some users may not have cars (a good assumption for some student populations), it is assumed that other Zipcar users may be families with one car whose Zipcar use can address competing demands for automobile access. Reservations are made on the Internet, and members use a proximity card to unlock and operate a specific car at the time agreed. The user picks up and returns the car to its dedicated parking space, so there are no trips to a rental office, and all financial transactions and approvals are embedded in the membership and reservation systems.

Under the agreement with Zipcar, Tufts has established dedicated parking for the Zipcars, and members of the university community who are over twenty-one are eligible to become Zipcar members at a reduced fee. Anecdotal evidence indicates that the Zipcars are popular among graduate students, faculty, and staff, and the program is being evaluated more systematically. The Zipcars in place at the university include two electric vehicles (EVs), provided to Tufts at no cost by Toyota. Electric vehicles, when properly charged and discharged, generate fewer emissions per mile than conventional automobiles; however, it is not currently clear whether the Zipcars displace automobile use (which would reduce emissions) or whether they result in new travel (which would increase emissions). Although this concept holds theoretical potential to reduce emissions, its potential is diminished significantly for institutions with exclusively undergraduate populations, because the number of drivers over twenty-one will be limited. Wellesley College has insured younger students in order to make Zipcars available to them.

One of the most time-consuming challenges associated with the initial phase of the Zipcar introduction was identifying dedicated spaces for parking the Zipcars. These cars require dedicated, reserved parking spaces so that they can be available for rental and have a space to return to. Despite the added parking capacity on the Medford campus associated with opening a parking garage in 1998, arriving at agreeable decisions on which spaces to "sacrifice" for the Zipcar minifleet (of two vehicles in Medford: one hybrid and one EV) proved extremely difficult. On our Boston campus, parking is even more difficult and each space can generate revenue, so the challenge was even greater.

Although TCI staff were aware that parking would have to be negotiated, the level of effort required exceeded worst-case assumptions. The reluctance to establish dedicated parking was not associated with negative views of Zipcars or the EV technology. In fact, the vehicles attracted positive attention and campus press shortly after delivery and before plug-in stations were established. Once the vehicles were charged and operating around campus, accolades poured in from both users and observers. The problem was that dedicated parking was seen as a privilege without precedent. Parking on campus is allocated on a first-come, first-served basis within zones (some zones are reserved for residents and others for faculty, staff, and commuting students), and the Zipcar needs challenged this long-standing system.

The technology of reserving cars online and providing access to them through a card that is activated remotely has been perfected by Zipcar. We believe that this technology can be employed on campuses to help manage university-owned vehicles that have multiple drivers (e.g., student-activity vans). This system may be able to improve vehicle tracking and reduce abuse—thereby saving modest driving miles and their associated emissions.

Reducing Emissions from University-Owned Vehicles

Many vehicles travel primarily within the campus so that staff can deliver mail, serve catered meals, or undertake facilities and grounds maintenance. A college or university may also own vans, small buses, or a fleet of buses. Tufts' fleet of over one hundred vehicles includes passenger cars, light-duty trucks, pickup trucks, vans, forklifts, police and safety vehicles, grounds maintenance, farm equipment, and construction equipment. Increasing the efficient use of these vehicles and planning for increased efficiency of future vehicles is an important strategy.

At Tufts, our experience shows that people who drive and manage university vehicles provide us with diverse settings for trying new technologies. For example, when Tufts bought a Toyota Prius, a compact hybrid drive car that runs on electricity and gasoline, TCI staff felt that public safety (police) was a logical application. The Tufts police spend a great deal of time at idle or slow speeds in cars that routinely get 18 miles per gallon. Results of student-led interviews indicated that the police were reluctant to embrace the Prius because of legitimate concerns about its ability to accommodate the lights, radios, and other police-related equipment. However, we were able to turn this setback into an opportunity by identifying others who drove university-owned vehicles. We found that the grounds manager drove an old police cruiser, which got 18 miles to the gallon, and would be happy to drive the Prius. His car, adorned with the TCI logo and a slogan that says "Tufts Division of Operations Saves Energy," is now well known around campus. (See figure 8.2.) The Crown Victoria was retired, and as a consequence, heat-trapping gas emissions were reduced by an estimated 6,000 pounds of CO_2 per year and fuel costs were reduced by $600 annually (before gas prices rose). The very positive reaction to the Prius has opened the door for additional demonstration projects, including the recent purchase of

Figure 8.2
The Tufts Prius

In addition to the RAV4 EVs, Tufts has purchased an electric tractor mower and has received a donation of a GEM, a General Motors electric golf-cart-style vehicle, that we use at our veterinary school.

The electric tractor has great promise for lawn mowing in a campus setting. Tufts has purchased one on a pilot basis, and initial feedback is positive. Electric mowers have the benefit of decreasing on-campus noise as well as emissions. At Tufts, our Electric Ox™ is used on the lower end of campus, coincidentally for mowing the organic-turf baseball field. At one point the grounds manager suggested that we install PV panels to offset the power for the mower to complete the picture!

Biodiesel Biodiesel is a fuel made from renewable sources that can be produced domestically. It is essentially vegetable oil that meets the ASTM D6751 industry standard as well as EPA emission standards. It can be used alone (commonly known as B100) or as a blend with diesel fuel. The most common blend is B20, which is 80 percent petroleum-based and 20 percent biodiesel.[21] The municipal government of Medford, Massachusetts, has used biodiesel at the city-operated cemetery since the fall of 2002 with no reported problems, although the diesel in the mix was a winter blend. When the city crews began to use it in their vehicles, the only reported difference was that the fuel filters on the engines had to be changed more frequently, initially because the B20 had the effect of cleaning the engine and removing deposits that had built up from the use of standard diesel.

Harvard University has been using B20 to fuel all of its diesel vehicles since 2003.[22] This includes the fleet of shuttle buses that circulate around Harvard's campus, a densely populated residential area in Cambridge, Massachusetts. The reduced particulate emissions have met with widespread support throughout the community. Additionally, the university continued to use the fuel during an extremely cold period in January 2004, one of the coldest months on record.[23] The biodiesel performed with no problems. This cold-weather performance should help dispel one of the persistent perceived problems with the fuel: that it would gel and not flow properly on cold days (it should be noted, however, that this is still an issue with pure biodiesel, B100).

During use, biodiesel reduces emission of unburned hydrocarbons, particulate matter, and carbon monoxide, while nitrogen oxides may be either decreased or increased depending on the duty cycle of the engine. Sulfur oxides are also significantly decreased. The USDA and DOE concluded that life-cycle emissions that contribute to global warming are reduced up to 78.5 percent compared to standard diesel if B100 is used; however, not all life-cycle assessments come to such optimistic conclusions on reductions in heat-trapping gas. If greenhouse gas reduction is the primary goal of introducing biodiesel, we recommend a careful review of the literature. Biodiesel is nontoxic; table salt is ten times more toxic. This makes it much easier to deal with if there is an accidental spill.[24] There is a premium cost associated with biodiesel, but it is small relative to the costs associated with some other emission reductions.

Compressed Natural Gas Vehicles The Massachusetts Bay Transportation Authority, the transit authority that serves Boston, purchased compressed natural gas (CNG)–powered buses at a premium approaching $50,000 per bus to meet the air quality mandates of the federal government and local activists. It is possible that they could have met these mandates through the use of emission-controlled diesel buses that utilize a blend of low-sulfur diesel and biodiesel at a much lower overall cost.

There is debate, however, as to the effectiveness of CNG in reducing greenhouse gas emissions. In some CNG-powered engines, unburned fuel escapes into the atmosphere. For this reason, Tufts decided not to pursue the use of CNG in police vehicles.

Travel Related to University Business

Faculty and staff at all levels of the organization travel on university business, and the emissions associated with their travel contribute to the buildup of greenhouse gases. Students also may travel for research collaboration, conference participation, and fieldwork. Depending on the type of accounting system in place in an organization, tracking the type of travel and the modal split may be challenging. For organizations with centralized travel services or travel-approval systems, records may be sufficient to capture a large portion of the travel related to university business.

There may be targeted opportunities to address university-related travel in departments that have staff in a few locations and who do a great deal of travel. At Tufts (and probably at most institutions), the university advancement (development) group accounts for considerable travel, with staff members frequently traveling off campus to meet with prospective donors. For automobile trips, may of these staff use their personal vehicles and receive reimbursement, or they may rent vehicles, depending on circumstances. This approach is costly to the university and generates considerable carbon emissions, especially if the vehicles are large. TCI initially hoped to use the shared-vehicle program run by Zipcar to determine whether the university might experience cost reductions if the advancement department had regular access to shared and efficient vehicles. Unlike vehicles intended for on-campus use, the range of the EV in the Zipcar fleet becomes a factor in an individual's decision on whether or not to use it for a particular trip, but the inclusion of a hybrid in the shared-vehicle pool solves that problem. Another factor that deterred this effort was the desire for staff to begin or end their travel from home, rather than returning to Tufts. A change in the department's office location (far from the shared vehicles) cut short our experiment, but we believe that it is worth considering elsewhere.

Air Travel Air travel accounts for a significant portion of Tufts' transportation-related emissions because planes (including jets) burn an enormous amount of fuel. For example, according to the *New York Times*, a New York–to–London round-trip flight in a Boeing 747 at 78 percent occupancy results in 2,776 pounds of CO_2 per person (round trip for the 747 generates 440 tons of CO_2).[25] Note that by comparison, the total annual emissions for the Tufts shuttle, a conventional diesel, are 70 tons per year! Early work suggests that university-related air travel may be the largest portion of the university's secondary greenhouse gas emissions.[26]

Given the cost (including time and risk potential as well as dollars) associated with travel, we assume that decision makers throughout the organization think carefully about whether making a trip is necessary to achieve their objectives. Unlike the computer-use situation (discussed in chapter 10), where we know people could turn off their computers but do not, we do not view business travel as being ripe for behavioral change. Having said that, the fact remains that each time someone at the university steps on a plane, some form of emission reducing action is needed to slow the growth of climate-altering gas emissions in the atmosphere.

For organizations that lack their own emission accounting and offset programs, the *New York Times* reports that third parties will plant trees or install energy-efficiency measures in public schools as a means of offsetting greenhouse gas emissions from air travel.[27] A number of organizations are emerging to provide these offsets (see the discussion of offsets later in this chapter).

Air travel for university business presents both great potential (because the associated emissions are so great) and some challenges. The greatest challenges may be educational in the sense that many people may be unaware of the emission impact of air travel, or they may be more complex: whether or not people are aware, they feel that their travel is necessary, although a carbon charge will also affect tight budgets. As noted earlier, we assume that the travel is necessary, but here are some suggestions for helping people test that assumption:

• Encourage faculty whose frequent flying is associated with speaking engagements to request both an honorarium and an emission offset. The dollars associated with the offset can be calculated using a generic calculator (such as carbonfund.org), establishing a flat fee, or using institution-specific figures. The advocate will have to decide on the best approach for the organization, and will have to establish a system to ensure follow-through.

• Ask whether videoconferencing is an option. Videolinks are being used for regularly scheduled classes and seminars that draw students (and/or faculty) from multiple campuses, and this is an area that deserves further exploration as we seek deeper reductions in emissions from transportation.

• When possible, take the train. Trains produce much lower carbon emissions per mile traveled than flying. Driving is also less carbon intensive on a per person per mile traveled.

Deliveries Deliveries to the university encompass a wide range of products and their frequency varies over the academic calendar. Although some document delivery is by bicycle messenger, the majority of deliveries are in trucks and cars of varying sizes. Vendors whose presence on campus is related to repairs and maintenance or consulting services also come in vehicles.

As with travel related to university business, it is assumed that deliveries are made on an as-needed basis, and there is no presumption on the part of TCI that this is an area ripe for reduction in emissions. That having been said, it is worth pointing out that there may be a relationship between deliveries and an organization's procurement practices. If procurement is centralized, it is conceivable that vehicle trips to and from campus will be reduced.

Which organization is responsible for counting the emissions associated with deliveries—the company delivering the product or the customer purchasing it? This is an inventory issue that needs to be discussed and resolved as increasing numbers of organizations take climate change seriously. Assumptions related to credit for emissions of this type should be noted in the greenhouse gas inventory.

Solid Waste and Greenhouse Gas Emissions

The life cycle of almost all products generates greenhouse gases in their use of raw materials, manufacture, transport, recycling, or disposal (see table 8.2). For this reason, using only needed products and managing their end use or disposal can help reduce emissions, even outside of the institution. Waste prevention and recycling can reduce methane from landfills, reduce emissions from incinerators, reduce energy consumption, and increase storage of carbon in trees.

Once a product has been used, it can be recycled into new products. While manufacturing products from recycled inputs still requires energy, fewer raw materials are necessary. Heat-trapping gas emissions are therefore offset by the avoided fossil-fuel use for raw material acquisition. In addition, for products that require wood or paper inputs, recycling reduces the need to cut down trees, increasing carbon sequestration in forests. If a product is not recycled at the end of its useful life, it goes through one of three waste management options: composting, combustion, or landfilling. All three use energy for transporting and managing the waste, and they produce additional greenhouse gases to varying degrees.

Recycling provides numerous benefits to an institution, and this is especially true for heat-trapping gas emissions. Recycling reduces the amount of energy used to make a product when compared to the amount

Table 8.2
Climate change effects of solid waste disposal

Activity	Material	Effect
Composting	Organic material such as food scraps and yard waste	Some carbon dioxide is released, but some carbon contained in organic matter is returned to soil and not released. There is a net reduction of heat-trapping gases.
Combustion	Solid waste	Combustion releases both carbon dioxide and nitrous oxide (a greenhouse gas that is 310 times more potent than carbon dioxide). When energy from combustion is used to generate electricity it offsets emissions from burning fossil fuels.
Landfilling	Solid waste	As material decomposes anaerobically, methane is released. (Methane is twenty-one times more heat trapping than carbon dioxide.) When methane from landfills is captured and used to generate electricity, greenhouse gas emissions are reduced. In a properly managed landfill, landfilling can serve as a long-term carbon sink for organic materials. Methane from landfills can be reduced by disposing of organic materials in alternatives to landfill.
Recycling	Paper, steel, aluminum, glass	Reduced energy for raw materials and manufacture results from recycling. Recycling also leads to a reduction in some noncarbon greenhouse gases (e.g., perfluorocarbons) for some materials such as aluminum and steel. The amount of carbon sequestered in forests will increase when paper is source reduced or recycled (because wood harvests will be reduced).

of energy necessary to produce that same product from virgin materials. From a purely financial perspective, recycling can also make sense, especially for a medium or large institution. By recycling, the institution can help reduce the solid waste disposal fees that it would otherwise pay to its waste hauler.

University climate action should include a recycling program. EPA provides an online tool for calculating the emission reductions that result from various waste reduction strategies.[28] Perhaps more importantly, our experience at Tufts suggests that a strong recycling program is critical to developing an awareness of environmental efforts on campus. In addition, university purchasing programs that support the purchase of recycled goods help to close the loop and support the market for recycled waste products.

At Tufts an end-of-year collection program addresses the "reuse" portion of the "reduce, reuse, recycle" hierarchy, reducing the amount of waste destined for disposal at the end of the year—when students leave residence halls—by more than 300 tons. This program, called Jumbo Drop after Tufts' elephant mascot Jumbo, is modeled on the Dump and Run program started by Lisa Heller.[29] The core idea is that collection bins are provided to students moving out, and materials are sorted for a variety of destinations, including donation to charity and resale. The volume of waste generated at the end-of-school-year move-out is simply enormous. It may be tempting to conclude that students are exhibiting behaviors associated with a wasteful society, and while this may be true in varying degrees, there are other explanations for the massive move-out waste. A junior who is spending his next year in Paris probably would be foolish to store possessions on or near campus. So this student is going to give away or discard many of his low-cost bulky items. The waste problem is exacerbated when this student's parents have driven 500 miles in their compact car to help him move out and take their son and his belongings home for the summer. Items that typically fall by the wayside in a scenario of this type include furniture, storage crates, laundry racks, foam mattress pads, area rugs or carpets, and shoe racks. Of course, there are many other scenarios and more diverse abandoned goods.

In an attempt to divert these goods from disposal, Tufts provides clearly labeled collection bins at the main exits of residence halls and encourages departing students to leave serviceable items and unopened nonperishable food. Even on move-out day some of the materials are diverted from the waste stream as departing students and their families treat the bins as a "give-and-take" area. Younger siblings are delighted to find speakers for their sound system, and frugal moms find space for a few good hangers in the family car.

After move-out, Tufts hires students, whose $50 per night extended-stay fee is waived, to help sort through the goods. Unopened nonperishable foods are donated to an organization serving the homeless. Foam mattress covers typically do not sell well and are donated to the animal rescue league, whose staff cut the foam and use it for bedding. Fans and other electronics are tested to see if they work and carpets are vacuumed. The "good stuff" is stored over the summer; schools use trailers or their hockey rink or some other large space. Then in the fall, tents are set up on a field, and incoming and returning students (as well as local community members) are invited to a giant yard sale. We call it the Jumbo Yard Sale. Our experience is that sales are directly influenced by the amount and quality of advertising for the event and weather. Goods left after the yard sale are donated or sent for disposal.

Climate Action in Dining Services

University dining halls, kitchens, and dish rooms are full of opportunities for reducing energy use and greenhouse gas emissions. These opportunities are greatest when a kitchen, servery, and dining room are designed or renovated, or when new equipment is purchased. As with any construction, care should be taken to select the most efficient equipment possible. Key opportunities exist in air-conditioning, exhaust, makeup air, refrigeration, dishwashing, water conservation (particularly of hot water), and cooking equipment. Some options may be improved by careful layout that groups equipment needing little or no exhaust together and equipment that needs more exhaust elsewhere. Lighting and HVAC in the dining areas as well as controls to shut equipment, exhaust, and lighting off are opportunities.

Dining-services facilities can participate in EPA's Energy Star program and can receive technical assistance and information on a range of energy-saving appliances, including commercial fryers, hot-food-holding cabinets, refrigerators, and freezers.[30] Other equipment such as heat-recovery systems that pull waste heat from display cases may be able to heat hot water with minimal additional energy. And infrared dish washers and fryers also offer significant operational savings compared to their conventional counterparts.

Kitchens and dining facilities that are in operation offer opportunities for conservation: turning off ovens, steam trays, unused refrigerators, or lights between uses, during breaks, or other times when not in use. Once equipment is off, ventilation systems can be turned off to save significant energy. Maintenance of equipment, such as cleaning coils, door seals, and lights, can also improve efficiency. Refrigerator and freezer controls can make a difference as well, and a university should select a maintenance contractor that has an understanding of ways to maximize efficiency.

As in other university facilities, replacing lighting, upgrading heating systems, and improving the efficiency of cooling systems are also critical. Energy controls can save money, too, since there are often large blocks of time between meals when the dining halls are unoccupied or have low occupancy.

In addition to having many environmental and health benefits, local and organic food also reduces greenhouse gas emissions from trucking and fertilizer applications. Vegetarian options also have lower greenhouse gas emissions per serving than meat because animals produce methane, a potent greenhouse gas, and there are greater life-cycle emissions associated with meat production.

Other Opportunities to Reduce Emissions

An institution's sources of greenhouse gas emissions other than carbon may vary widely in type, quantity, and relationship to the core activities of the organization. Although it is difficult to generalize, our experience suggests that it is very important to identify other sources because they may be much more potent than carbon dioxide in their impact on climate change. The following is a brief discussion of other sources that may be present at many universities, depending on their mix of programs. As the list shows, attention to climate action is needed in many academic departments, laboratories, and clinics. In some cases it will be relatively easy to change practices, but in others, alternative approaches may be difficult to identify.

While the quantities of some of these sources of heat-trapping gases may seem small, many have significant global warming potential (see chapter 2 and appendix A). For example, methane's global warming potential (GWP) is 21 times more powerful than carbon dioxide, nitrous oxide's GWP is 310, and many refrigerants have GWPs that are 1,000 to 5,000 times more potent than carbon dioxide.

Management of University Forests

University forests often are separate from the main campus and are used for biology and forestry education and research. Because trees and other vegetation can sequester carbon from the atmosphere, management of these forestlands can help to offset or store carbon generated elsewhere by the institution. Use of university forests as a source of wood products can also store carbon for a relatively long time, rather than letting it decay back into the atmosphere. In addition to managing their own forests, universities with forestry programs can be a source of knowledge to help others learn how a range of forestry practices affect climate change.

Managers of university forests should look to maximizing carbon storage capacity of existing forested lands, enhancing the long-term potential to sequester carbon in existing forests through increases in productivity, and addressing fire management and pest control. Because wood can sequester carbon for long periods, university policy should consider promoting the use of wood products.

Livestock

The normal digestive process of most animals produces methane. While true for everything from termites to humans, emission reduction efforts focus on commercially valuable species. Cattle, buffalo, sheep, goats, and camels are the major sources.[31] Tufts' School of Veterinary Medicine maintains a research herd at the Grafton campus. In 2001, the herd consisted of approximately 50 dairy cows, more than 300 swine, and a few sheep, goats, and horses. The herd's emissions of methane add up despite the small herd size because methane has a global warming potential 21 times greater than carbon dioxide. Agricultural and veterinary schools may have a significant portion of their emissions from their herds and may have opportunities to reduce these emissions by altering feedstocks, changing feeding schedules, and improving the activity and health of the animals. A wide variety of techniques and management practices are currently implemented to various degrees among U.S. livestock producers; these techniques improve production efficiency and reduce methane emissions per unit of product produced. More widespread use of these techniques, as well as the implementation of new techniques, will enable further reductions of methane emissions from livestock.[32] University research related to reducing emissions from herds may be of considerable interest worldwide as society places increasing priority on climate change.

Manure Management When livestock manure is handled under anaerobic conditions (in an oxygen-free environment), microbial fermentation of the waste produces methane. Liquid and slurry waste management systems are common manure management techniques and are especially conducive to methane production. Because confined livestock operations such as dairy and hog farms rely on liquid and/or slurry systems to manage a large portion of their manure, they account for a majority of all animal manure methane emissions in the United States.[33]

Feasible and cost-effective technologies exist to recover methane produced from the liquid manure management systems used at large dairy and swine operations. Because methane can be used as a fuel, methane gas recovered by any of the available methods provides a renewable energy source, often used to generate heat or electricity or to operate chillers for milk refrigeration. Methane emissions from livestock manure can also be reduced by using aerobic treatment, such as spreading manure on fields as fertilizer. However, this must be done with care to avoid oversaturation and water quality problems.

Fertilizer

Industrial fertilizers and organic materials such as manure add nitrogen to soils. Any nitrogen not fully utilized by the crops grown in these soils undergoes natural chemical and biological transformations that can produce nitrous oxide (N_2O), a highly potent greenhouse gas. There is some uncertainty about how to best manage fertilizers, but universities that use large-scale fertilization programs for campus greens, athletic fields, or agriculture may be able to reduce climate change impacts by modifying practices. Techniques that include changing the timing, amounts, and fertilizer type are among the possible solutions. As noted earlier, Tufts has established an organic-turf baseball field.

Tillage

Because nitrogen is stored in soils, universities with large agricultural lands may be able to reduce greenhouse gas emissions by changing their soil-tilling practices. This change is being advocated by corporate members of the Chicago Climate Exchange, who are paying farmers in the Midwest to use alternative techniques that decrease the release of emissions. Universities with large agricultural holdings may want to investigate this approach in research and in action.

Refrigerants

Many refrigerants used in air conditioners and refrigerators are potent greenhouse gases. Some of these gases are also ozone-depleting substances, and thus are regulated by the federal Clean Air Act. Strict compliance with refrigerant management programs will both ensure compliance and reduce emissions. However, many substitutes for ozone-depleting substances are global warming gases. To make matters more difficult, alternatives may reduce equipment efficiency. This issue remains difficult and in need of new technology. In the meantime, refrigerant management to prevent leakage is critical.

Telecommunications

As discussed in chapter 7, the proliferation of central computer facilities and communication equipment and of the dedicated spaces for this equipment increases demand for electricity, both to power the equipment and to provide climate control to the spaces. Universities should look for ways to purchase equipment that can withstand a wider range of temperature swings and to ensure that communications storage spaces are not overconditioned, oversized, or otherwise inefficient.

Tree Planting

Strategic planting of trees and shrubbery on campuses can reduce energy demands and increase the sequestration of carbon. Landscaping with trees can provide shade and reduce wind speeds and thus reduce energy needs related to heating and cooling. Because half the dry weight of wood is carbon, as trees add mass to trunks, limbs, and roots, carbon is stored in relatively long-lived structures instead of being released to the atmosphere. Thus, programs to support tree planting can help reduce greenhouse gas emissions in a variety of ways. It is estimated that urban (or campus) tree planting is five to ten times as effective in reducing global warming as a tree planted in the wild because it lowers demand for air-conditioning.

Heat-Trapping Gases in Medicine and Medical School

Nitrous oxide is a potent greenhouse gas with a warming potential 310 times greater than that of carbon dioxide over a 100-year time frame. This gas is used in clinics in Tufts School of Dental Medicine and other dental schools as an anesthetic (laughing gas), and because of its potency, even small quantities are a concern. Scavenging equipment is used because nitrous oxide has deleterious effects on worker health; however, the equipment still allows some releases to the environment. Sulfur hexafluoride is an even more potent heat-trapping gas. It is commonly used in eye surgery, and while it is used in very small quantities, its release is of concern.

Carbon Offsets and Trading

Market mechanisms for encouraging climate change action or for mitigating climate change through financial transactions are emerging on a largely voluntary basis, especially in light of the absence of federal policy. Carbon offsets and carbon trading are two such mechanisms.

Carbon Emission Offsets A project that reduces greenhouse gas emissions outside of the university to compensate for the university's own emissions is called an emission offset. Often the institution simply pays a fee per ton of carbon dioxide so that someone else in another location can implement an energy-efficiency measure, switch fuels, or plant trees in an amount calculated to offset the university's emissions.

Organizations are available to help colleges and universities offset their carbon through a variety of means. One such organization is the Climate Trust. The Trust helped Lewis and Clark College students offset emissions so that the college could reach the Kyoto goal. The premise of these offsets is that it often cheaper and easier to implement emission reduction measures elsewhere than to undertake them at the institution. Offsets may involve tree planting in developing countries or purchasing energy-efficient heating systems in primary and secondary schools.

Because climate change is a global problem, offsets can be anywhere; however, it is important that investments in offsets be verifiable. It is also critical to ensure that the offsets are additional—that is, that they would not have happened without the funds from the college or university purchasing the offset. Furthermore, it is important to understand how offsets affect emissions; the ownership of the carbon must be transferred to avoid double counting.

At Tufts we considered becoming a source of carbon offsets, but found that our projects were not really suitable because benefits are long term and diffuse, as compared to major fuel-switching opportunities in industrial locations.

Emission Trading Programs Emission trading programs allow private entities to buy and sell pollution reductions that are achieved. These market-based systems present opportunities for reducing aggregate pollution levels at a lower cost to society than if rules required everyone to reduce by the same amount. An emission cap and trading system is currently used in the United States to reduce sulfur dioxide, a pollutant that contributes to acid deposition. These programs provide incentives to polluters to reduce emissions and sell credits to others whose reductions may be more costly. This is an approach that could be applied to carbon dioxide emissions as well, either domestically or internationally.

The Chicago Climate Exchange® (CCX®) is the first effort to develop a carbon trading mechanism in the United States. The exchange is a greenhouse gas emission reduction and trading pilot program for emission sources and offset projects in the United States, Canada, and Mexico. Projects also include Brazil. CCX® is a self regulatory, rules-based exchange designed and governed by CCX® members. The members have made a voluntary, legally binding commitment to reduce their emissions of greenhouse gases by 4 percent below the average of their 1998–2001 baseline by 2006, the last year of the pilot program.

Most members of CCX® are large corporations. In 2003 Tufts became the first university member of the exchange. Other university members include the Universities of Oklahoma, Minnesota, and Iowa. Tufts has found that the annual reporting requirements are time consuming, and we continue to evaluate our participation. We recognize that emissions trading programs will likely be part of the national and regional policy solutions of the future; this is the basis for the emerging trading system in the Northeast and Mid-Atlantic States called the Regional Greenhouse Gas Initiative.

Summary

Throughout the university there are numerous opportunities to take climate action. Often these opportunities have multiple benefits and their implementation yields surprising results. The challenge for a university climate change program is to understand those actions that are most appropriate for their program and institution. These decisions will be based on the magnitude of the opportunity to reduce emissions, the transferability of the project, its educational value, and the willingness and availability of relevant university personnel to work on the effort. Finding allies from inside and outside the institution's walls and asking the market to deliver products and services with lower emissions can have multiple benefits on and off campus.

9

Planning and Policies for Climate Change

Having a plan for campuswide emission reduction is an essential part of climate action; however, it is most likely to be effective if it is viewed as part of a larger whole. Colleges and universities that successfully embed climate change in the full range of their planning and policies are more likely to realize comprehensive and effective climate action. If institutions are located in jurisdictions where government planning and policies also give priority to climate change, municipal or state actions may increase the effectiveness of college or university efforts. For example, a college located in an area that is expanding its public transit network will have more nonautomobile options for faculty, student, and staff travel. Or a university located in a country or state that is aggressively pursuing wind-power development will be able to purchase green electricity at competitive rates. Even with supportive government actions, planning and policies will be needed to achieve campus climate action goals.

This chapter opens with a discussion of planning for the climate action program, and then examines several college and university plans and policies that are related to climate action. Our experience suggests that a climate action program cannot be effective unless it actively informs other planning efforts on campus and in turn is informed by other plans. Managing these complex relationships effectively is a challenge. Like so much of climate change action, climate change planning is an iterative process that involves learning by doing.

Steps to Climate Action Planning

Across the country municipalities, states, and corporations are creating climate change action plans to identify feasible and effective policies to reduce their heat-trapping gas emissions. A university climate action plan can to the same. As with any good program plan, the climate change action plan will be part of a dynamic, evolving process in which goals are set, measurable objectives are identified, progress is evaluated, and modifications are made as implementation experience is gained, and as external factors change. To be most useful, the university climate action plan should include short- and long-term goals as well as a set of objectives. The plan should also be informed by other university plans such as the master plan, facilities plans, and financial plans.

When the Tufts Climate Initiative began in 1998, a climate action plan was developed. At that time, the plan was used to determine whether meeting the Kyoto goal was reasonable and achievable and to provide some direction on next steps. But the initial plan was lofty and vague. Over the next five years our planning efforts at were guided by a clear focus on short-term projects that are likely to succeed and toward larger and longer-term projects. Because we are grant funded, our annual requests for funding serve as our primary planning documents, but each year the goal has remained the same. Box 9.1 summarizes the major elements of the Tufts Greenhouse Gas Reduction Plan.

Information on state and municipal action plan development can be useful in formulating a campus action plan.[1] In the following paragraphs, we offer basic steps for developing climate action plans and programs that draw on our own knowledge and experience.

Prepare a Campus Greenhouse Gas Inventory

The inventory is the baseline against which subsequent progress is measured. See chapter 3 and appendix C for details on how to develop the campus inventory. Inventory data will also help developers of the climate action program describe the problems to be solved. For example, after conducting an inventory, you might conclude that "our electricity is the major source of greenhouse gas emissions because it is procured from a utility that burns coal." An information-rich problem description will be useful in communicating with a wide range of campus decision makers and will help establish priorities. This is also a good time to conduct research on programs being implemented on other campuses to learn about their successes and failures, and to survey the literature more broadly for insights that can inform program decisions. In program

Box 9.1
Tufts Climate Initiative Greenhouse Gas Reduction Plan

Goals
1. Reduce university emissions consistent with the goals of the Kyoto Protocol and the New England Governors.
2. Increase awareness of climate change and actions to reduce greenhouse gas emissions.

Focus Areas
1. Improve energy efficiency of new buildings.
2. Improve energy efficiency of existing buildings.
3. Engage Tufts community in personal actions to reduce emissions.
4. Explore alternative fuels and fuel switching.
5. Research new technologies and monitor success.
6. Educate about climate change and climate change action: within Tufts and external to Tufts.

Implementation Strategies
1. Work with university personnel and support their efforts.
2. Take action where we can be successful.
3. Work on strategies that build institutional knowledge and existing action.
4. Work with investment that is underway.

development terms, this set of activities is usually conducted at the needs assessment stage.

Establish Program Goals

Goals are ambitious and long term. We discuss different climate action goals in this chapter and also in chapter 3. Goals generally fall into two large categories, process and outcome. A process goal related to climate change might be "to integrate climate change into the curriculum of every department in the college," and an outcome goal might be "to generate no net emissions of greenhouse gas (become climate neutral)." Process goals describe how you are going to do it, and outcome goals explain the changes you want to achieve.

Articulate a Program Theory

The program theory does not have to be a grand theory comparable to the theory of relativity. And different elements of the program will have different theories. For a program to encourage students to turn off their computers when not in use, the program theory can be as simple as the following: "People don't turn off their computers at night because they don't understand that electricity from fossil fuel increases heat-trapping gas emissions. If we teach people that electricity use contributes to climate change, then people will turn off their computers when they are not in use." The reason for elaborating a theory is that it is essential to evaluating the success of a program.

Establish Measurable Objectives

For both program goals and process goals, identify a set of objectives for each. Objectives should be framed in terms that are understood by decision makers, and should have dates and quantifiable expectations. For example, a measurable objective for the process goal of integrating climate change into the curriculum is "incorporate climate change into two courses in the biology department and one course in the English department by 2008." A measurable objective for the outcome goal is "decrease greenhouse gas emissions from electricity 1 percent per year beginning in 2009."

Identify Program Components

Program components are the specific activities you will undertake. They should be defined in a way that makes sense on your campus. For instance, three components could be: installing lighting controls in the biology building, replacing all incandescent chandelier bulbs in the chapel, and upgrading the boiler in the engineering building. You might have few components at the beginning of a climate action program and you can add components as the program gains resources. For example, a program component related to the goal of integrating climate change into the classroom might be a faculty development program of the type we describe in chapter 11. Another might be a peer training program such as our Eco-Reps, described in chapter 10, who are given information on the relationship between turning off computers and lights and generating heat-trapping gases, and who in turn convey this information to others in their residences. A program component related to the goal of reducing emissions might be replacing the athletic department van with a new, very fuel efficient vehicle.

Evaluate Progress Regularly

In Fran Jacobs's five-tiered approach to program evaluation,[2] there is an explicit recognition that programs follow a developmental path in which early efforts tend to produce results that are related to process and subsequent efforts tend to produce outcome results. This is reasonable because changes often take time to manifest themselves. In this example, progress in the early years can and should be evaluated in terms of number of faculty participating in development programs, number of departments incorporating climate change in their curricula, and so forth. But after a while, it will be necessary to evaluate progress in terms of outcomes. Let's say for the sake of argument that five years have passed, all the faculty on campus have gone through a development program on climate change, climate change has been integrated fully into the curriculum, and there is an active and popular peer training program in which students talk with one another about the relationship between turning off computers and reducing emissions.

An evaluation is conducted, and annual process objectives are met or exceeded, yet the updated inventory shows that just as much electricity is being used each year on campus as the day the climate action program started. In this case, we would say that the education components of the program were implemented as planned, so we would look for other explanations for the failure to reduce electricity use. One place to look is the theory behind the program. That is, perhaps education alone—emphasizing that leaving computers on generates emissions—does not cause people to change their behavior. And in fact, there is evidence in the literature on social marketing indicating that knowledge of this type may be insufficient to yield changed behavior. If only a more thorough literature review had been done during program development!

Other explanations are also possible. One important source of illumination is the inventory, and another is the other planning and implementation activities on campus. Perhaps two new residential buildings were added and the campus now houses 150 more students that it did when the climate action program started. If this were the case, individuals on campus could have turned off their computers more frequently, but there were more people on campus. Careful analysis of the inventory data and knowledge of other campus activities should reveal exactly what occurred. Whether decreased per capita consumption of electricity (same electricity use, but 150 more residential students) should be declared a program success or a program failure is a matter for the climate action program to consider as objectives are modified for the next round of implementation. One thing is certain in this example: the outcome goal of no net greenhouse gas emissions will not be achieved if the components of the program only relate to personal actions such as turning off computers at night.

This hypothetical example is intended to illustrate basic steps in program planning and development and to emphasize the value of frequent program evaluation, even at the early stages of the climate action program. Equally important, the example is designed to emphasize that a climate action program both informs and is informed by other planning and development activities on campus.

Here's an example of an evaluation at Tufts. In a spring 2005 course on climate change, an exchange student from Germany undertook an evaluation of the emission reductions taken during renovations of a small residence called the French House or Schmalz House project. The full description of the initial project is in chapter 5. In the evaluation, the student's primary interest was examining electricity- and fuel-use data and comparing it with prerenovation consumption levels. He was given a tour of the house by a current resident, and was careful to document emission reducing features for his report. We were disappointed to learn from him that the Energy Star refrigerator Tufts installed to reduce emissions has acquired a mate. The second fridge is also an efficient model, but having two was not in the original plan. Depending on the age and other attributes of the equipment, two Energy Star fridges could use more electricity than one conventional unit.

Then the student delivered more bad news: the energy-efficient front-loading washing machine had been replaced with a conventional top-loading model. Who took the washer? And why? Adding to the intrigue is the fact that the initial positive experience with the front-loader at Schmalz House contributed to a decision to procure front-loaders for other residences. In an ironic twist, Schmalz House might be the only residence on campus with a top-loading washer.

The bottom line: despite the additional fridge and the top-loading washer, electricity use in the house generated 6,086 fewer pounds of CO_2 in 2004 than in 1998, the year before emission reducing renovations.

Total emissions in 2004 were half of those in 1998. The greatest share of reductions came from switching from oil to gas heat. And further: the evaluation pointed out that the university's system for managing the house is not climate-sensitive. Many decision makers are involved, but not all have energy efficiency as the only priority.

In the next several sections we offer specific examples of the complex relationship between the climate action program and other campus planning and development activities.

University Master Planning and Climate Action

Major decisions of the university are informed by plans, many of which have heat-trapping emissions implications. The university administration and trustees undertake long-term, large-scale planning; however, more detailed and short-term planning takes place in all academic and administrative departments. A comprehensive discussion of planning at academic institutions is beyond the scope of this book, but we will explore intersections between university planning and climate change action.

University planning takes many forms, from annual planning to long-term physical facility planning. Short-term planning is often manifested in budget decisions, while long-range planning looks at program development and physical plant expansion and modification as well as strategic planning for teaching, research, and scholarship. A campus build-out, identifying potential building locations, is one outcome of a traditional long-range campus physical planning process. Climate change action in university planning includes giving priority to energy decisions, evaluating energy implications of alternative plans, and focusing on systems reliability. It also includes thinking about the legacy of the institution, its fundraising strategies, its investments, and its teaching and research programs.

Setting Climate Goals

Perhaps one of the most important phases of university planning is the goal setting that sets the stage for detailed plans. These goals are critical for comprehensive action on climate change. Goals for addressing climate change have been embraced by governments, industry, and individuals (see chapter 2). For example, countries that have ratified the

Kyoto Protocol and states and provinces that have endorsed the New England Governors and Eastern Canadian Premiers climate action statements also set goals for climate change. In Burlington, Vermont, and in an increasing number of towns, cities, and campuses, individuals are signing up for the 10% Challenge, a pledge to reduce individual emissions by 10 percent.[3]

Goal statements are powerful motivators that provide direction to planners and implementers. They send signals to individuals inside and outside the organization that an issue is important. Over and over college and university climate advocates ask us, "How do we secure top management commitment?" And while the answer to that question is not clear or simple, the frequency of the question indicates that people understand the value of a statement of high-level commitment.

Planning goals for climate change can range from a comprehensive goal of specific greenhouse gas reductions (such as those embraced by Tufts and other colleges and universities) to a commitment to certain actions such as green power purchase, high-efficiency building construction, or leadership measures. Goals may also include the importance of climate change to the teaching of science, policy, political science, or other disciplines. This is not a far-fetched idea—many colleges and universities are doing just this in their master planning process. Commitments framed as objectives are also useful and, because they are more specific and easily understood and measured, they may be more effective at motivating changes related to climate change and energy efficiency. Objectives implement established goals and are more detailed. For example, many universities are pledging to make buildings "green" by meeting green building ranking criteria. Others are purchasing renewable energy or carbon offsets. Table 9.1 provides some additional examples of goals.

Campus Trends and Emissions

Across the country universities are increasing electricity use. In the late 1990s the demand started growing at about 1 percent a year on our Medford campus and about 3 percent a year universitywide, as students have brought more and more electrical appliances to campus. These devices range from cell phones to air purifiers and average twelve to fifteen per residential student. Faculty and staff also have dramatically

Table 9.1
Goals related to climate change

Institutions	Goal
MIT	MIT's standards include "conserve energy, seeking continuous reductions in our per capita energy consumption." This is done through using LEED Silver as their intermediate platform that all buildings are built to. They have added a "Plus" of their own design and intend to revisit these standards in the short term to determine if they make sense for the long term.[1]
The Ivy Council (representing all of the Ivies)	The Ivies have resolved to build all new construction to the LEED Silver standard or higher and set a goal of 15% of energy from renewable sources by 2010.[2]
Bowdoin College	Bowdoin's recent Master Plan includes a commitment to LEED standards for new buildings, including two new dorms being built.[3]

1. http://web.mit.edu/environment/commitment/env_goals.html.
2. http://www.ivycouncil.org/pdf/ivycorps_resolution_11-15-2003.pdf.
3. www.bowdoin.edu/bowdoinmagazine/archives/features/002903.shtml.

increased the number of computers, printers, scanners, and other types of equipment in their offices and laboratories. Model results for a new residence hall at Tufts showed that the building may use more steam heat during unoccupied times than during occupied periods, simply because during winter vacations the heat-producing electric equipment will be turned off and steam heat may have to be increased to keep pipes from freezing.

Colleges and universities are also increasing the amenities on campus by equipping residence halls with high-end appliances and providing additional services such as theaters and other recreational facilities. See box 9.2 for specific examples. University planners may overlook the energy, maintenance, and infrastructure implications of these trends. Climate change action will need to address the implications of these facilities in light of university goals and the perceived market demand for luxury.

Revising Master Plans to Address Climate-Altering Emissions

A campus master plan is a wonderful opportunity to articulate an inspirational vision for the university and to move these ideas from concepts

Box 9.2
Campus amenities

This material is excerpted from an article appearing in the *New York Times* in October 2003.

Whether evident in student unions, recreational centers or residence halls (please, do not call them dorms) the competition for students is yielding amenities once unimaginable on college campuses, spurring a national debate over the difference between educational necessity and excess.

Critics call them multimillion-dollar luxuries that are driving up university debts and inflating the cost of education. Colleges defend them as compulsory attractions in the scramble for top students and faculty, ignored at their own institutional peril. And somewhere in the middle sit those who have only one analogy for the building boom taking place.

"An arms race," said Clare Cotton, president of the Association of Independent Colleges and Universities in Massachusetts. "It's exactly the psychology of an arms race. From the outside it seems totally crazy, but from the inside it feels necessary and compelling."

Students now get massages, pedicures and manicures at the University of Wisconsin in Oshkosh, while Washington State University boasts of having the largest Jacuzzi on the West Coast. It holds 53 people.

Play one of 52 golf courses from around the world on the room-sized golf simulators at Indiana University of Pennsylvania—which use real balls and clubs.

Only about 100 miles away, Pennsylvania State University's student center has two ballrooms, three art galleries, a movie theater with surround sound and a 200-gallon tropical ecosystem with newts and salamanders. Oh, and a separate 550-gallon salt-water aquarium with a live coral reef.

Ohio State University is spending $140 million to build what its peers enviously refer to as the Taj Mahal, a 657,000-square-foot complex featuring kayaks and canoes, indoor batting cages and ropes courses, massages and a climbing wall big enough for 50 students to scale simultaneously. On the drawing board at the University of Southern Mississippi are plans for a full-fledged water park, complete with water slides, a meandering river and something called a wet deck—a flat, moving sheet of water so that students can lie back and stay cool while sunbathing.

Source: Greg Winter, "Jacuzzi U? A Battle of Perks to Lure Students," *New York Times*, October 5, 2003, 1.

into practice—including, for example, active citizenship and sustainability. Many colleges are expanding their campuses by adding new buildings, or undertaking major renovations or additions to existing buildings. Typically these efforts are carefully orchestrated by a campus plan that looks well into the future. Rarely, however, does this plan consider either the consequences of the campus growth for infrastructure and its effect on the institution's net climate-altering emissions.

While climate change planning may not yet be common and some might argue is beyond the purview of master plans, we think that integrating these two long-range planning efforts is precisely the type of action that leading universities and other organizations will take. There are important and logical ways that capital planning efforts can intersect with climate action. First, considering climate change action will emphasize issues of energy efficiency, building siting, building reuse, and fuel sources. These activities have implications for the campus emission profile. Second, climate change models indicate that storm, flood, and other weather conditions are likely to change dramatically in some areas. Colleges and universities that have planned for these changing conditions will be better equipped to handle them. These activities fall into the category of adaptations and typically will not reduce emissions, but they may help protect people and property.

Adapting to Changes Caused by Global Warming

Across the United States and beyond, the effects of climate change may be dramatic. The United Nations Environment Programme estimates that worldwide economic losses due to natural disasters appear to be doubling every ten years, and the next decade will reach $150 billion a year.[4] Natural disasters appear to be more frequent and more severe.[5] Campus planners should consider the energy delivery, increased frequency of floods, and consequences of higher average temperatures that are predicted (often by academics) to occur. Box 9.3 shows some of these effects.

Some members of the private sector are taking climate-related warnings seriously. In particular, the insurance and reinsurance industries are actively working to bring attention to this problem. In some cases, insurance companies are canceling policies for coastal properties due to the increasing risk of storm-related flooding. Swiss Re, the world's second-largest reinsurance company, believes that losses in their industry could

Box 9.3
Effects of climate change and planning implications

Energy demands While less heating fuel may be needed in winter, the summer peak electricity load may increase.
Energy cost Climate change may cause energy costs to rise due to increased demand and increased frequency of disruption in supply.
Storms Increased storm activity could mean higher repair costs as well as increased service outages. Storm events may become more severe, so planning systems to handle additional stormwater will be needed.
Water Water supply and quality may be impacted by climate change. There could be issues of shortages or changes in the quality, especially if salinity changes in coastal areas.
Flooding Current infrastructure planning often considers the 100- or 150-year floodplain, but with increased extreme weather events and higher seas, these planning benchmarks need to be revised. University infrastructure involving sewage capacity, underground utilities, and transport should be considered.
Agriculture Food costs may be increased because of global warming changes throughout North America.
Sea-level rise Sea-level rise may be more than 3 meters in the next century. This rapid change will dramatically alter coasts and affect properties far inland as well.

be significant. Chris Walker, a top executive from the company, told an audience at Tufts in January 2003 that in Europe, it is a foregone conclusion that climate change will have an impact and that corporate America needs a wake-up call.[6]

The university's ability to secure insurance in the face of changing conditions is an aspect of planning that deserves attention. Insurance for flooding, storm damage (including wind, snow, and ice loads), and power disruption are important considerations for the future.

Predicting climate changes specific to a particular geographic region is an area of climate science that is not fully developed. This is due in large part to the great complexity of global systems, as well as to the limitations of models used to predict effects and the uncertainty associated with local variations in key parameters such as precipitation and evapotranspiration. A pioneering study conducted at Tufts called Climate Impacts on Metro Boston (CLIMB) was an early attempt to model the effects of climate change on a relatively small region.[7] The results show that the Boston area will likely experience temperature- and storm-related changes that can affect campus infrastructure. These predicted changes should influence campus master planning, facilities planning, and emergency planning. For example, predicted warmer temperatures will result in a greater need for air-conditioning to maintain comfort levels. At present, Tufts does not routinely air-condition residence halls, an approach that may become unrealistic if temperature increases occur sooner than scientists originally predicted. If more campus spaces are air-conditioned, the operational costs and maintenance considerations may favor central cooling rather than the current system of room air-conditioning units in many spaces. Schools in other parts of the country face these and other planning challenges associated with climate change.

For many areas, and the colleges and universities within them, the effect of sea-level rise may be devastating. As the earth's surface warms, global sea levels will rise due to melting glaciers, thermal expansion of the oceans, and changes to the major Greenland and Antarctic ice sheets. In the near term (the next fifty years) these effects may cause increased flooding, salt contamination of freshwater supplies, and exacerbated storm action. Colleges and universities near oceans should consider their vulnerability to these devastating effects.

Linking Adaptation and Mitigation Strategies

While master plans should create strategies for adapting to the effects of climate change, they are also central to implementing the actions to reduce emissions as well as to finding ways to connect both adaptation and mitigation. For example, an increased reliance on distributed power or on-site generation from combined heat and power (see chapter 6), renewable energy, or alternative fuels may decrease heat-trapping gas emissions and make the university less vulnerable to power outages or storm events. Likewise, attention to energy efficiency can help to reduce the impact of rising costs.

Distributed generation systems, particularly those that rely on fuel cells, can be extremely efficient and can result in substantial reduction of heat-trapping gas emissions associated with powering the university. Cogeneration systems, in which both heat and electricity are generated by the same plant, may also be an excellent choice for many campuses. Clark University in Worcester, Massachusetts, is one institution that made an early investment in cogeneration and has benefited financially over the life cycle of the investment. A downside of distributed generation that needs to be taken into consideration in planning relates to backup power. If for any reason the on-site campus generation system is shutdown, for either routine maintenance or for equipment failure, purchasing grid power during shutdown may be extremely costly. This is one of several factors that must be taken into account in evaluating on-site generation systems.

Emergency Planning Planning for emergencies, especially those that may result in power outages, should be factored into the full cost of energy systems. Too often generators for backup power are tacked onto projects without any systematic thinking about their capacity to function in different types of emergencies.

An incident affecting our main campus in Medford helped clarify these issues for us. In the summer of 2002, a utility transmission line and its backup failed. The campus was without grid-connected power for two weeks. Diesel generators with a total capacity of 5 megawatts were brought in from all over the Northeast—from as far away as Maryland—and the campus community was urged to conserve as much as possible. Facilities staff worked day and night to connect the generators into the power feeds and to distribute the limited power appropriately throughout the campus.

Among the lessons we learned was how vulnerable we are when the power fails. The outage occurred during a slow time of year, when most students were not on campus, academic buildings and research facilities were not fully occupied, and demands for electricity and HVAC were not at a maximum. Had the outage occurred during the semester or in a winter storm, our ability to conserve would have been reduced and generators might have been less available and would therefore have been capable of addressing a smaller portion of demand. The cost of the outage was extremely high and although some of the direct dollars were borne by the utility, the staff time and lost productivity were not compensated. Though unrelated to climate change in any way, the power outage helped us realize the need to think differently about the vulnerability that can be associated with climate change.

The incident taught us that the definition of emergency power depends on the circumstances. For example, during a one- or two-hour power outage, emergency power is sufficient if it serves life safety needs. However, outages that have longer duration increase the sphere of those whose needs are critical. For instance, climate control in laboratories where long-term research is being carried out can become critical to the faculty and their departments. Even computer use can be critical if a faculty member is working on a proposal due that day and the only copy of the document resides on a desktop computer. Communications, voice mail, telephone service, and copy machines become critical for some departments when the duration of a power outage lasts more than a day.

The relationship between emergency energy planning and climate change has several dimensions that warrant consideration:

1. As climate change affects local weather events and storms become more powerful or frequent, interruptions to service may be more frequent. If this means that power outages will last for longer periods, the suitability of the campus emergency power system needs to be examined to determine whether critical needs can be met under different scenarios.
2. The cost of an emergency such as a power outage can be enormous. These costs can change the financial equation for evaluating different energy systems. For example, distributed energy systems using fuel cells, cogeneration, and other on-site generation systems generally have a higher first cost than grid-connected systems. However, in evaluating the vulnerability of the grid system, college or university planners may feel that the increased reliability of on-site generation compensates for the higher first cost.
3. Some solutions, particularly on-site cogeneration, can increase reliability and decrease emissions of heat trapping gases.

Other dimensions of emergency planning are worth considering as well. They include emergency communications within the campus and outside of the campus, transportation home, temporary housing, evacuations, and emergency medical access.

Components of a Climate-Sensitive Master Plan

Climate change can no longer be thought of as a problem of the future. Current campus infrastructure and any facilities being planned will be affected by global warming. Therefore, the campus master plan is an appropriate place to identify the ways that campus growth and change can be climate sensitive or may be affected by climate change. Some details on the opportunities for climate change and master planning follow.

The Master Plan Can Link Buildings with the University Mission A focus on climate change provides an opportunity to think about the links between the master plan, the physical plant, and the research and academic mission. This focus may include creating explicit demonstrations of technology, allowing students and faculty to participate in planning activities, evaluating the proposed plan, and creating plans that allow for visible rather than invisible campus infrastructure and energy systems. For instance, Brown is linking their master plan with their Academic Excellence Initiative.[8] This is not a new idea; in the fall of 1994, Middlebury's President John M. McCardell named environmental studies and awareness as a Peak of Excellence at Middlebury College. The Environmental Peak, along with five other peaks, defines a vision for the future of the college. "These Peaks," noted President McCardell, "are like the Green Mountains of Vermont: our vertebrae, the source of our strength and the definer of our character, and the reason why many people choose to come to Middlebury."[9]

The Master Plan Can Direct Growth and Planning Beyond just a plan for new buildings, landscapes, and travel patterns, the master plan can be a tool for making decisions. These decisions should include energy cost, energy reliability, and reduction of heat-trapping gases. For example, Johns Hopkins' plan provides similar guidance on related issues by describing a "process to develop a plan . . . that directs physical growth in a way that:

- Preserves and enhances existing natural systems.
- Improves the aesthetic character of the campus. . . .
- Strengthens the relationship of the campus with the surrounding neighborhoods."[10]

The Master Planning Process Can Build on the Vast Expertise and Ideas of the Community Members of the college or university community have expertise in a range of disciplines that are reflected in the planning process. Faculty have expertise in many relevant areas, including physical planning, land use, sustainable communities, environmental hazards, stormwater treatment, native landscapes, and energy systems.

Universities routinely provide opportunities for meaningful participation during the initial planning and throughout the implementation so that community expertise can be incorporated into the plan. For example, in an e-mail to the Harvard community, then-president Lawrence Summers described a December 2003 planning meeting of seventy faculty, staff, and students as well as series of task forces to begin the planning for the Allston campus. He welcomed "comments, ideas, and reactions."

Brandeis offers a web-based approach, including a campus master planning website with a suggestion box for specific comments on their process.[11]

The extent to which climate change, sustainability, and related issues are raised or articulated in a master plan may depend on the firm selected to help the institution develop the plan, as well as on the extent to which the university community is involved. In 2004, Tufts engaged a firm to produce a master plan, and the vision and goals articulated by the master plan steering committee resulted in a draft plan that had a central focus on identifying sites for new buildings. When the draft was circulated for comments, several faculty, staff, and students were disappointed that it did not emphasize a commitment to sustainability. This was not because plan developers viewed sustainability negatively; it was simply overlooked as a powerful way to strengthen the plan. Only after input from the community did planners include an important opportunity to link the university's educational mission and its signature programs with its evolving physical plant.

With others in the community, we acted as advocates for master plan revisions that captured opportunities to emphasize sustainability and energy efficiency. In deciding on an advocacy strategy, we elected to take an educational approach, advising our colleagues of missed opportunities in the draft plan. It is our experience that faculty and administrators are open to these issues, but often need to have the connections well articulated for them.

The Master Plan Can Include More Than Building Sites The master plan is an opportunity to integrate planning for community and campus utilities, connectivity, and energy systems. For example, "the scope of the University of Chicago Master Plan included an in-depth study of campus utilities and detailed building assessments for a representative group of university facilities."[12]

The Master Plan Can Address Energy Supply The university master plan will be strategic if it also addresses the institution's reliance on fossil fuels, the long-term supplies of these fuels, and any anticipated scarcity and/or price increases. While some regions may have more to worry about than others, there is agreement that worldwide oil reserves are finite and may be insufficient to meet demand within our lifetime. Some predict that scarcity may begin within the next ten years, affecting supplies and price.[13] All of these factors will create challenges and potential costs for institutions. Master plans can begin to address this vulnerability by including careful attention to energy infrastructure with a twenty- to fifty-year horizon.

Facilities Planning

Facilities planning is guided by the larger goals of the master plan and informed by short-term needs. Especially on older campuses, facilities planning may focus primarily on existing buildings and facilities, treating new construction separately. Good facilities planning is the integration of space and program improvement, campus modernization, deferred maintenance, and utility planning. As we describe in chapter 6, energy delivery, building use, and mechanical and electrical systems intersect with climate change in this area. Certainly efficient buildings, life-cycle costing, and maintenance issues have implications for climate change and for the implementation of the master plan goals. Because reducing energy use in existing buildings is critical to facilities planning, it is a logical place to give energy issues priority status.

Facilities managers often feel that they have insufficient resources to accomplish all that needs to be done. Furthermore, there is often a grave lack of understanding or appreciation in the campus community for the complex task of managing and maintaining hundreds of buildings and their supporting systems. For instance, most members of the community never give the condition of the roof or the maintenance of the heating system a thought until the roof leaks or their building is uncomfortably hot or cold. At the same time, there is often significant pressure for visible facilities work such as the renovation of existing spaces, the installation of technology, or an improved aesthetic. It is important for climate change advocates to understand these pressures and work with facilities managers to advocate for adequate budgets for energy upgrades and building maintenance.

Larry Goldstein, president of Campus Strategies, Inc., suggests that facilities managers advocate for facility expenditures in the same way that technology managers advocate for technology. He believes that technology is often amply funded with little understanding of what it will really do, based on a notion that technology will add value to the institution. In contrast, facilities investment—a manifestation of facilities planning—is sold and funded based on its return on investment. In addition, facilities managers can advocate for projects based on opportunity costs, action of competitive institutions, and the documented cost from not funding.[14]

Facilities planning for climate change must be holistic and realistic. It must address energy supply and demand, building use, controls, and maintenance.

Financial Planning

Financial planning can create opportunities to put university resources to work in numerous ways that can have emission reduction benefits. Because the allocation of budgets generally reflects priorities, financial plans can link operating and construction costs, consider life-cycle costs, and link budgets across departments. Each of these measures can help to create incentives for energy conservation and give priority to energy-related projects in a range of departments.

University advancement (fundraising) efforts can also benefit from innovative climate change action, green buildings, and related efforts. Advancement planning can be enhanced by identifying these efforts as "selling points" for increased donations from alumni or other donors. Some foundations, such as the Kresge Foundation, are beginning to encourage green buildings as a part of their funding process.

Investment Planning for Colleges and Universities

A great deal of interest surrounds college and university investments as endowments have waxed and waned with the U.S. stock market, and as members of the university community periodically raise concerns about the ethics and practices of companies in which institutions hold shares. Climate change creates both vulnerability and opportunities for institutional investors, and careful planning can increase the likelihood of positive outcomes.

As we mentioned in chapter 5, a 2002 report from CERES titled *Value at Risk: Climate Change and the Future of Governance* examines the implications of climate change for company directors, fund managers, and trustees and concludes that they will be abdicating their fiduciary responsibility if they fail to take climate change into account in evaluating investments.[15] The United Nations Environment Programme's Financial Initiative also concludes that all areas of investment will be affected by climate change.[16] We view this as a compelling argument for colleges and universities to scrutinize their endowment portfolios and consider divestment from companies whose core businesses are large-scale generators of climate-altering gases, such as utilities relying on coal.

Endowed institutions will need to think broadly about their portfolio exposure to climate change risk. According to *Value at Risk*, cement and semiconductor manufacturers, along with those that have methane emissions from livestock, marine operations, aviation, or aluminum production, among others, will create significantly higher risk for portfolios because of their greenhouse gas emissions.[17] Faculty or outside experts familiar with heat-trapping gas emission intensities of different manufacturing operations can assist endowment officers in identifying particularly vulnerable investments.

The European Union adopted a mandatory carbon trading scheme that began operation in January 2005.[18] This trading scheme could have a direct impact on the value of college and university portfolios depending on the extent to which they are invested in European companies. This trading scheme will also affect many American companies that have European operations. This is only one example of the financial exposure that universities and other endowed institutions must examine as the climate change issue shifts from a scientific pursuit focused on problem definition to actions that have impacts on all levels of university operations and decision making.

On the positive side, climate change may offer institutions new investment opportunities that can potentially meet financial, environmental, and social goals. Leading companies are taking climate change seriously and are reducing emissions from their facilities, products, and processes at the same time that they increase their share value.[19]

Members of the college or university community frequently pressure the administration to divest shares in companies whose practices they find unacceptable. With respect to climate change, an alternative to divestment is shareholder activism, in which the college or university might choose to communicate with the companies vulnerable to climate change to determine whether they can shift to a more sustainable business model. CERES advocates for this particular approach for universities. CERES recommends that universities, at the very least, vote their proxies instead of automatically giving their votes to management.[20] By convincing companies that climate action is important to universities and other institutional investors, this activism encourages companies to change their methods and activities.

Universities may also want to consider investments in energy systems that create predictable energy prices. Long-term investments in wind farms or other alternative energy sources may provide a known price for power that emits a great deal less heat-trapping gas than conventional sources. Executing long-term contracts for fuels such as #2 and #6 oil has historically been a common aspect of energy management at colleges and universities. As prices have become more upwardly volatile, companies providing fuel oil have offered shorter-term contracts that may result in extra effort on the part of the energy manager and unpredictably large costs to the college. In 2006, Tufts experienced an increase of several million dollars in its energy bill, prompting a memorandum to faculty from the administration discussing our budget crunch.[21] If an investment in green power offers predictability in costs as well as an opportunity to demonstrate social responsibility, we argue it is worth serious consideration.

Using Endowment Funds to Create an Energy Account That Is Paid Back from Savings

Another potential investment opportunity for a portion of the university's endowment money may be in campus efficiency upgrades. While this idea may be new to investment officials, operations specialists have

known for years that many efficiency projects can offer rates of return that are much better than the typical investment portfolio. Joe Romm's work suggests that in American businesses, many projects can have a 3.1-year payback.[22] While university buildings may have somewhat slower payback potential because of their rates of use, constant renovations over the years, and centralized systems, energy projects still show tremendous promise as investment opportunities. The rates of return from efficiency upgrades are essentially guaranteed because, with proper planning, the institution can know, at a minimum, how much money can be saved on energy bills. This type of investment makes sense for a school that is investing its money over a long time horizon and is looking to diversify its investments. The low-risk, high-reward payoff available through efficiency should be very appealing to university business officers, especially considering the financial difficulties many endowment managers have faced over the past several years with volatile share prices for publicly traded companies. The efficiency investment continues to pay over time and has the added benefit of helping the institution to meet environmental and community relations goals. The returns on these investments can be treated like any other return on the investment portfolio, "compounded" into new high-return energy investments, or designated for a special purpose to illustrate the benefits of energy savings to the campus community.

Educational Planning

There are several ways that educational planning can be used to influence the campus emission profile, many of which are unique to each college. For institutions that periodically select overarching themes for the community, climate change is a choice that may motivate faculty to modify course offerings, students to change decisions on turning off their computers, and administrators and staff to reevaluate vulnerability of campus infrastructure. For instance, on campuses with an endowed speaker program, selecting climate change as a theme can yield multiple benefits. A speaker series allows the college to attract a national figure to address the community, and is used by faculty in many departments as an opportunity to attract additional speakers and teach classes on related subject matter. We hope that opportunities of this type may result

in longer-term effects such as a seminar on climate change or disciplinary or interdisciplinary courses addressing select aspects of climate change. For example, a course in marine biology might expand the discussion on coral reefs to explore what is known and not known about the effects of climate change on the Great Barrier Reef. We take the view that climate change is so important to education that there is a place for its consideration in many aspects of academic life. We address climate change in the classroom in detail in chapter 11, where our primary focus is on strategies to use the college as a learning laboratory.

Policies to Reduce Emissions

University policies can facilitate emission reduction; sometimes existing policies will need climate-sensitive modification and in other cases, new links between policy decisions and emissions will be forged.

Academic Calendar

In response to the energy crisis of the 1970s, many colleges and universities modified their academic schedules. In the Northeast, a common strategy was to schedule a longer break in the winter between the fall and spring semesters. Over break, buildings such as student residences and classrooms could be heated to a minimum (to prevent freezing pipes) in the December-January period. This approach was designed to save money by reducing demand during a period when campus energy use is typically high.

Colleges and universities could consider schedule modifications if climate change begins to affect historical patterns of energy use. For example, colleges and universities in the Southeast would take a rather different approach from the example cited above. If increases in temperature and humidity result in significantly more electricity use for air-conditioning in fall, one solution for Southeastern colleges and universities is to start classes well after Labor Day, and extend the academic year into May, and/or shorten winter and spring breaks. Along similar lines, institutions located in coastal areas susceptible to hurricanes and tropical storms should consider schedule modifications if climate changes result in increased frequency of events that can be life threatening. The fall 2004 string of hurricanes and tropical storms resulted in costly and disruptive evacuations by colleges and universities in Southeast coastal areas. Worse, in fall 2005, Hurricane Katrina hit many universities on and near the Gulf Coast just as students were arriving for the new academic year. Beginning the academic year later in the fall, when the hurricane season ends, is an adaptation for which the benefits may outweigh the costs.

Dramatic schedule modifications at colleges will not be undertaken lightly because they have cascading effects on athletic leagues and competitions, family vacation schedules, textbook ordering, and planning for hotel and event space in surrounding communities. On the other hand, it is one of many adaptations to climate change that decision makers may need to exercise.

Space Utilization and Scheduling Policies

The least expensive way to gain additional classroom and meeting spaces is to adjust the class schedule so that building spaces are used for the maximum number of hours. Using spaces that already exist rather than building new buildings is a climate-friendly strategy as well. Space utilization rates (the percent of time a space is used) at colleges and universities vary widely, and in some institutions with lower rates, there may be opportunities to increase available space by changing how courses (and their necessary meeting or laboratory spaces) are held and increasing the intensity of their space utilization. Space planners can:

- Ensure that all departments and schools use the same schedule (same periods)
- Ensure that classes are being scheduled in the less desirable times (early morning, Fridays, and even Saturdays)
- Consider running laboratory classes throughout the day, rather than just in the afternoons

At Tufts, these strategies effectively increased available classrooms wired for technology by 20 percent without significant dollars having to be spent for capital investment.

Building Use and Performance Policies

Policies are important instruments for affecting decisions in the college community; however, the extent to which policies are effective or consistent with the institution's culture varies widely. Because academia often honors and rewards independent thinking and creativity, policies that mandate equipment or behavior may be few and countercultural, although practices vary across institutions. Exceptions occur when legal requirements dictate action (such as in the handling and disposal of hazardous materials) or when life safety is paramount. Policies that can favor climate change include thermal comfort policies (heating and cooling), building use policies, and purchasing and travel policies. In the absence of government regulation on carbon emissions, it may be difficult to catalyze climate action policies and planning at colleges and universities, particularly if it is perceived that policies constrain the range of choices available to faculty, students, or staff.

There are challenges and opportunities associated with a variety of policies that can influence energy consumption and carbon dioxide emissions. We offer ideas about energy-related policies that may inspire institutions to craft policies that are uniquely suited to their circumstances and culture. Since managing building energy use is critical to achieving climate-altering gas reductions, policies that address how buildings are used and the expectations for building conditions are critical.

Target Temperature Policy Modifying indoor temperature targets can save energy. Temperature policies can create expectations for building occupants. Policies serve as a benchmark that will help to identify problems, mitigate complaints, and, in theory, reduce the use of energy-hogging equipment such as space heaters. A temperature policy should determine appropriate temperature set points (summer and winter) for the university from industry standards (such as those of the American Society of Heating, Refrigerating and Air-Conditioning Engineers or ASHRAE) and determine the process for creating exceptions. Many schools—including Dartmouth, SUNY Buffalo, and Middlebury—have taken this approach.

Temperature policies are tricky. Enforcement is difficult. Furthermore, on campuses with aging systems or historic buildings, it may be difficult to deliver and maintain stable temperatures. Lack of uniformity across a campus or even within a building can also be a problem. Some special cases—such as a student's or staff member's health or special research conditions—may require exceptions, so a policy should provide a process for addressing these needs.

Temperature policies are especially challenging in an era of "customer service." A policy can protect facilities personnel from having to please everyone all the time. On some campuses, a pocket-sized laminated printout of the temperature policy is given to facilities personnel, who can respond to complaints by testing temperatures and discussing the results with occupants in light of the official and readily available policy guidelines. Developing an explicit complaint process and training key facilities personnel about how to respond to complaints will help with successful policy implementation.

Temperature policies can also help to reduce oversizing in new construction by addressing target temperatures under extreme conditions. For example, a summer cooling policy may indicate a target temperature of 78°. However, it may provide for higher indoor temperatures when outside temperatures exceed 95°. In the Boston area, these extremely hot days occur rarely, and most commonly in August when the university is the least utilized. Allowing higher target temperatures during those conditions will allow smaller, more efficient air-conditioning systems in new and existing buildings.

Weekend and Night Setback Policy (Winter and Summer) Academic buildings on many campuses are rarely used during nights and weekends. Some buildings may be unoccupied or may have a very low occupancy (60 percent of the week's total hours). Significantly reducing building temperatures in the heating season or increasing building temperatures in the cooling season during periods of low occupancy (nights and weekends) can save energy. Typically, buildings can be assessed on an individual basis, and accommodations will be made for scheduled special events such as evening performances at the theater.

As we discussed above, this policy requires determining appropriate temperature set points for the university from industry standards and determining the process for creating exceptions. It also requires creating an expectation and understanding among faculty and staff regarding evening and weekend workplace conditions. As with target temperatures, there are difficulties in implementing this effort, but the savings can be substantial.

A weekend and evening setback policy will need careful monitoring to ensure that no unintended additional energy use occurs. For example, if faculty or staff use space heaters and air conditioners excessively during off hours, the net energy consumption may increase.

Electric Space Heater Policy Electric space heaters are generally an inefficient way to heat a space. However, in some cases, they are used as supplemental heat because the central systems are inadequate or because the distribution of heat within a building is uneven. In other cases, space heaters may be used to raise temperatures to levels determined to be comfortable by an individual, in excess of a temperature set by policy and delivered by central systems. Space heater technology varies, and some models provide greater efficiency and safety than others.

Space heater policy must be linked to the temperature policies described above. At Tufts we have discussed (but not implemented) a voluntary registration of space heaters as a means of determining where central systems may need to be improved. Where space heaters are justified, central facilities would provide the units, purchasing the most efficient equipment on the market. Another possibility is to make space heaters available to individuals working nights and weekends when temperatures may be cooler due to setbacks. Borrowing approved space heaters from campus police is an option for implementing this system for night and weekend use.

Since space heaters can be readily purchased and are often brought from home, enforcement is difficult. Effective management of this policy will require clear communication of its importance for cost containment and safety as well as buy-in from department managers. At the same time, delivery of target temperatures by central systems is essential so that the root cause of space heater proliferation is addressed.

Window Air-Conditioner Policy Window air conditioners are generally an inefficient way to provide cooling. In many climates, window fans will meet comfort needs under most, but not all, conditions. At the same time, efficiency gains can result from purchasing or leasing the most efficient equipment possible and from selecting the smallest air-conditioning unit for the room being cooled. When window air conditioners are necessary, removing or sealing the units can increase comfort during the winter and reduce heating demand. Air-conditioning units contain regulated refrigerants, all of which are heat-trapping gases. Handling these regulated materials adds extra university responsibility.

At Tufts, the discussion of air-conditioning policy is further complicated by the existing process for budgeting and paying for air conditioners. If a department occupies a building with central air-conditioning, air-conditioning systems, maintenance, and energy costs are covered in the university's overhead so the service is provided outside of the department budget. If window units are used, the department must pay to purchase or lease the units as well as maintain them. However, departments do not pay the resulting operational costs because electricity is paid from the central overhead account.

The window air-conditioning policy is also closely linked to temperature policy, to setback policy, and to individual building performance. In some cases, window units that service only one office and are shut off when unoccupied can be more efficient over time than central systems that condition hallways and common spaces and run continuously. Careful thought and policy development in this area is critical, but will need to address university-specific building use and climate conditions.

The first steps are to create a review process for air conditioners and to determine the need in light of temperature policy. The policy for air conditioners may include:

- Restricting access in some applications
- Centralizing purchasing or contracting to maximize efficiency (e.g., requiring purchase of Energy Star units) and ensure proper sizing
- Creating centralized systems for installations and winterization
- Linking operating cost to users
- Providing guidance about use (e.g., turn off during nights and weekends)

Other Policies with Climate Implications

There are many other policy areas that can link university operations and decision making to climate change action. In addition to achieving energy efficiency and achieving cost savings, these policies are also opportunities to underscore the university's commitment to the issues.

Purchasing and contract policies are opportunities to specify efficiency since many energy-using machines are available with a range of efficiency. These include copy machines, printers, computers, refrigerators,

10

Personal Action Initiatives

Personal decisions are vitally important to the climate action effort because they cumulatively account for a significant and growing portion of electricity use on a residential campus. Equally important, a focus on personal action allows us to expand the educational reach of our commitment to reducing greenhouse gases. Our experience and that of others who have undertaken a wide range of comparable efforts suggests that motivating individuals to make choices in favor of the environment is not easy.

Knowledge and Action

Studies indicate that the general understanding of climate change is so inaccurate that even if people were motivated to change behaviors, they would not know what actions to take. In 2001, Steven Brechin found that only 15 percent of Americans surveyed correctly identified fossil-fuel burning as the primary cause of climate change.[1] More recently, John D. Sterman and Linda Booth Sweeney created a series of tasks to explore peoples' intuitive understanding of climate change. Focusing on students at MIT, Harvard, and the Graduate School of Business at the University of Chicago, they concluded that "highly educated people have extremely poor understanding of global warming."[2] One of the phenomena that Sterman and Sweeney explored was the relationship between emission rates and atmospheric concentrations of heat-trapping gas. In the hypothetical examples given, study subjects did not understand that even if emissions were immediately to drop to zero, concentrations of climate-altering gases would continue to rise for several decades before leveling off. This demonstration of the subjects' failure to understand delay in the climate system suggests a substantial challenge for all of us as educators and communicators. Scientists understand that if we wait to reduce emissions, we have a much larger problem; politicians do not understand the price of delay. In a similar test, Sterman and Sweeney found that widely held mental models of climate change violate the principle of conservation of matter.[3]

Over a decade ago, W. Kempton, J. S. Boster, and J. A. Hartley used anthropological methods to understand how Americans from a wide range of professional and political backgrounds understand climate change.[4] Their assessment reveals that, with few exceptions, the people interviewed and surveyed had a mental model of climate change that is completely different from the explanation offered by scientists. Indeed, they found that the model of stratospheric ozone depletion was most often offered as an explanation of climate change. With this flawed model, it would not be possible for people to associate their use of cars, electricity, and home heating with climate change problems. Although the study by Kempton and colleagues was undertaken years ago, and there is reason to think a similar study done today would yield different results, people may retain flawed mental models.

The Cost of Misconceptions

Many approaches can be used to calculate the cost of misconceptions related to global warming. For example, it is possible to estimate the cost of one or many events related to extreme weather, an approach that has inspired the insurance industry to advocate for prompt climate action. In a report to Congress weeks after Hurricane Katrina, the Congressional Research Service estimated the private-insurer losses at $40 to $60 billion, making Hurricane Katrina the costliest single event in U.S. history, exceeding even the terrorist attacks of September 11, 2001. The report calculates total economic losses (insured and uninsured) at over $200 billion.[5]

Sea-level rise that results in coastal flooding will engulf significant cultural and historical places and structures, and will cause flooding of sites where generations of people have lived and are buried. Loss of land has economic, emotional, and intergenerational costs. Some costs associated with global warming seem inestimable even though dollar figures can be attached.

Decisions made today have climate-related costs because they foreclose other options for a decade or more. If an electric utility company invests today in a new coal-fired power plant, it will produce large amounts of carbon dioxide for its entire thirty- to fifty-year useful life. If that facility powers your college or university, it will be difficult to meet ambitious climate action goals unless electricity is purchased from another vendor. If a gas-guzzling SUV is manufactured today, it will produce excessive carbon dioxide emissions over its ten- to fifteen-year life.

Inspire Understanding of Climate Systems, Then Provide Facts about Global Warming

Scientific research organizations, intergovernmental agencies, and governments are developing materials that illustrate the climate system and provide facts about climate change. Many are available on the Internet and reflect different levels of complexity for diverse audiences. Just a few of these resources include the United Nations, whose UNEP/GRID-Arendal website includes slides and an interactive model;[6] the Hadley Centre;[7] and the Environmental Protection Agency.[8] Teaching modules related to climate change and personal action also are available on the Internet. Appendix B contains a list of resources.

Knowledge Does not Equal Action

Campaigns conducted by environmental activists often assume that education is essential to behavior change. This assumption is inherently attractive for those of us at academic institutions; however, there is compelling evidence that a great deal more than understanding is needed to inspire people to change their habits. Recently we convened a group of climate change professionals at Tufts, and asked them whether their personal emissions from their own activities were greater than or less than in 1990. The majority of participants, all of whom understood the urgency of climate change, replied that their own emissions and certainly the collective emissions of the assembled group were greater now than in 1990. That is discouraging. But the problem is even more complex.

The Nature of Personal Choices

In the course of our daily activities, we make several choices that have climate change implications. Some decisions have a great deal more impact than others, and they vary in relevance to different demographic groups. Wealth may be a factor in an individual's emission profile. For example, the residential transportation energy consumption survey available from the Department of Energy shows that higher income is correlated with more vehicle miles traveled.[9] Emissions from personal travel usually account for a significant portion of our personal inventories. As with institutions, another significant portion of personal emissions comes from home heating and cooling and from electricity use. Secondary impacts result from air travel as well as purchasing and food choices. Table 10.1 presents a qualitative grouping of selected actions and impacts where time is an important variable. A phenomenon usually discussed in the context of countries is technology lock-in, but it also applies to individuals. This is the idea that once you purchase a refrigerator you will continue to use it until it expires (usually for fifteen years or more),

Table 10.1
Relative groupings of personal actions and emission impacts

Greater impact over long term • Live in a small space • Use efficient heating, lighting, and cooling in living space • Set back thermostats • Insulate living space • Use solar hot water and PV • Vote for individuals and policies that will make emission reductions a priority • Use fuel-efficient vehicle if driving is necessary	*Modest impact over long term* • Use energy-efficient appliances • Use energy-saving settings on computer • Use rechargeable batteries and solar charger for portable music devices • Recycle as much as possible, and recycle all aluminum • Minimize hot water use • Live near work, stores, or transit and walk or bike as often as possible • Eat lower on the food chain
Greater impact over short term • Minimize air travel for work and vacation	*Modest impact over short term* • Purchase locally grown food in season • Unplug when on vacation (many appliances use electricity even when "off" because they are in the ready mode)

even though much more efficient models may come on the market. The other side of this issue is technology lockout. When you purchase a new gas-fired hot water heater, it means that you may have locked out using a non-fossil-fuel alternative such as solar thermal for at least the water heater's warranty period (usually ten years) and probably until the unit fails.

The impact of many of these actions can be quantified using assumptions or historical-use data (utility bills) if they are available. These calculations can also be used to determine payback periods for long-term capital investments such as efficient vehicles and appliances and solar hot water.

Should an effort be made quantify emission reductions from personal actions on campus? It depends on how the information is to be used. For campuswide climate action planning and strategy development, it may be sufficient to track changes in electricity and heating use and make estimates about the implications for personal action. On the other hand, in the context of a course or workshop on climate change, it can be extremely valuable to go through the calculations, particularly if the learning objectives relate to awareness raising, data gathering, using emission factors, converting units, and exercising common sense. Individuals can use web-based tools such as EPA's Personal Greenhouse Gas Calculator[10] to establish a rough understanding of their emission profile and of the reductions associated with select measures such as driving a more efficient car or driving fewer miles. In a graduate course we ask students to use web-based tools to calculate their emissions and develop an action plan for emission reduction that they can really live with. Several said that their an emission and financial problem was the roommate who kept cranking up the heat when no one was looking.

Staff and Faculty Choices

We believe that there is a link between peoples' choices at home and those at work. TCI holds seminars for faculty and staff on saving energy at home. In the publicity for these events, we emphasize that many energy-saving measures can also save money. On a campus where sessions of this type are being introduced for the first time, it would be very interesting to launch this in the form of an experiment. This could be done by an individual faculty member or in conjunction with a course.

Through baseline surveys and careful follow-up, it should be possible to learn whether and how new emission reducing actions are implemented at home and then are transferred to work. We know that there are few incentives for individual emission reduction actions on campus, particularly because individual departments and research units are not billed directly for energy costs (or savings). Our thought is that informational sessions with employees can help shape community norms at the same time that they provide information on how individuals can save money at home with energy-efficiency measures. Testing these ideas formally can add to knowledge in the field and also can inform the effectiveness of emission reduction efforts on campus.

Student Choices

In fall 1999, a survey was conducted as part of an undergraduate political science class at Tufts. In a course on survey design and analysis, students were asked to take a representative sample of undergraduates, and to pose questions related to global citizenship and the environment ($n = 288$).

Of the respondents, 74 percent agreed or strongly agreed with the statement "The US government should take an active role in the global effort to curb the problem of rapid climate change." And 67 percent said that overall, environmental issues were moderately important or very important to them. Findings related to the importance of environmental issues are not surprising, because they compare favorably with national surveys[11] as well as with surveys of recent Tufts graduates.

The survey then elicited information about personal habits that have an impact on the environment. When asked "Approximately how often are the lights left on in your room or work area in your house when it is not occupied?", 32 percent responded nearly always or sometimes and 68 percent rarely or just about never. The next question was "And about how often do you leave your computer on when you are not using it?", to which 76 percent responded nearly always or sometimes and 22 percent responded rarely or just about never.

No explicit attempt was made in the student survey to determine whether respondents were aware of the link between their energy use and its contribution to climate change. However, anecdotal evidence suggests they are not. The chair of the Tufts student environmental organization was also active in the student corporation that rents "microfridges" to students for use in their individual rooms. In conversation, she revealed that she never questioned the energy efficiency of these devices or the significant energy saving that would result if students instead used the common refrigerators that the university provides in each residence.

The survey information on individual behavior was used by the Tufts Climate Initiative to inform its activities. The three personal action areas we identified as priorities were replacing incandescent bulbs with compact fluorescents, encouraging members of the campus community to turn off their computers for at least six hours in each twenty-four-hour period, and encouraging appliance shutoff and heating setbacks during vacation.

Plug Loads

Plug loads involve the electricity used by equipment installed and operated by building occupants. In a typical campus room, students have a storehouse of electrical appliances and devices, including desk lamps, electric razors, and lighted makeup mirrors. According to a March 2003 survey by Miami University, the average freshman takes eighteen appliances to campus.[12] Table 10.2 shows some of these devices along with the amount of electricity they require. We mentioned plug loads earlier because on our campus, the growth in plug loads is outstripping our efforts to conserve electricity from other sources such as lighting. Here we talk about plug loads in the context of personal action.

Once electricity-using equipment arrives on campus, the institution generally does not tell students and faculty how to use the equipment—and most often the climate change action advocate is relying on personal action to modify use. This can be a hard sell. It is our sense that many colleges and universities have been reluctant to address the trend of growing plug loads despite the fact that it is costing a great deal of money. Here is an example:

> In a renovation a few years ago, Wright State doubled to four the number of electrical outlets in each of the 162 rooms at Hamilton Hall, increased the number of circuit breakers, installed new electrical-switch gear and rewired fuse boxes and dorm rooms. The cost was about $500,000, or $1,000 per student.[13]

Table 10.2
Electronic devices

Electronic device	Approximate electricity demand (watts)	Type of use
Television	80	Hours/week
DVD player	19	Hours/week
Desktop computer	200	Hours/day to continuous
Laptop computer	45	Hours/day to continuous
Printer	54–120	Intermittent, standby constant
Fan	75–150	Hours/week
Chargers	Varies	Hours/day
Stereo	250	Hours/day
Alarm clock	5	Continuous
Electric toothbrush	2	Continuous
Lamps	40–100	Hours/day
Refrigerator (18 cu. ft.)	500	Continuous
Microwave	1,300	Minutes/day
Hair dryer	1,500	Minutes/day
Iron	600	Minutes/day

And that was just the cost for renovations. Wright State will also be paying each year for increased electricity use by students.

We note with interest that that Rowan University encourages students to bring oil-filled space heaters for winter use.[14] Indeed, there is evidence to suggest that institutions are facilitating increased energy use by providing a variety of amenities, including cable television hookups and built-in refrigerators, as ways of marketing the quality of campus life to potential students.[15] On the other hand, some electricity-using devices, such as halogen lamps, are prohibited in student residences. Bans that are widespread across many campuses are described in box 10.1.

Focus on Computers

Since the 1990s the number of computers on all campuses has increased. In the 1980s many offices had shared computers, and few students arrived on campus with a personal computer. Now most students bring

Box 10.1
Banned in Boston (and many other places)

Halogen lamps have been banned on most campuses. This was originally done for safety because of the fire risk; several fires occurred when posters or curtains touched the hot lamps and were ignited. The halogen lamps also use massive amounts of electricity (300–600 watts) and generate so much heat that additional building conditioning is required.

Personal *air conditioners* are banned on most campuses. Most universities do not give a reason, but it probably relates to their impact on peak load. Increasingly, students are receiving medical waivers for university-supplied air conditioners.

The limit on *refrigerator* size across several campuses is about 3.0 cubic feet. This appears related primarily to concern for physical size, but in general, the larger the unit, the more energy it is using (the exception is full-sized fridges, which paradoxically tend to be more efficient on a per square foot basis). Many campuses are now going to a rental model for individual refrigerators. This could be an untapped opportunity for savings because most of the small refrigerators rented are the cheapest available and are very inefficient. A change to reliance on Energy Star models for campus rentals could reduce energy consumption. Sharing full-size, very efficient refrigerators would be even better.

computers with them, nearly all faculty and staff members have dedicated computers, and computers are being used for an increasing array of academic and communications purposes. Library catalogs are computerized, and many journals are available only in electronic form. E-mail has become a reliable way of transferring data and text files, as well as for conducting communications of many types. The Internet has become a valuable tool for teaching and learning, and for gaining access to data. Among students, computer use goes well beyond basic academics and communications to include instant messaging, image storage and processing, audio (especially music) and video capability, web business development, and a host of other activities and functions.

As a result of these use patterns, computers are consuming electricity in residence halls, offices and laboratories, classrooms, and libraries all over campus around the clock. While some twenty-four-hour computer use is necessary and desirable for communications, energy management, and real-time data collection, certainly not all is necessary, and this is where personal action will play a role unless the university has in place

Table 10.3
Select comparisons of laptop and desktop energy use

	Standard (watts)	Sleep mode (watts)	Annual savings of sleep mode 10 hours/day (maximum) (kWh)	System off (watts)
Desktop CPU alone	160–250	3.3	1,000	1.4
CRT monitor	40–100	4.5	349	3.8
Flat-screen monitor	32–75	<2	266	<1
CPU and CRT	200–405	7.8	1,450	5.2
CPU and flat screen	192–380	<5.3	1,368	<2.4
Laptop	15–45	5	150	<1

an energy management system that powers down buildings when they are not occupied.

Addressing computer-use problems involves two elements. The first is an effort to have computers and computer-related equipment shut off when not in use. The second is to change the type of monitor and computer purchased. Table 10.3 makes comparisons of energy use for different types of computers (laptops and PCs) while awake, in sleep mode, and off. Cumulative energy savings from selecting laptops and flat screens and shutting off or enabling the sleep feature of the computer can be significant. Multiply those savings across a campus at current electricity rates and you can accumulate enough savings to award additional scholarships every year.

Cool the Climate! Power Off!

Not only has the number of electronic devices on campus increased, so have their hours of use. At Tufts we began addressing the use of electronics by creating simple messages. As part of this effort, we spent a great deal of time debunking the myths of computer and light shutdowns and we linked the messages to climate change. See box 10.2 for computer myths. These are modest programs; however, they are concrete actions that may start to make the connections for people, even if they do not yield widespread changes in practice. TCI prepared informational brochures explaining the link between climate change and electricity use, one focusing on lightbulbs and one on computer use. We have had many

Box 10.2
Computer myths

Myth #1: Turning off my computer is bad for my computer.
Fact: The Lawrence Berkeley National Laboratory states that modern hard disks are not affected by frequent shutdowns and that equipment may actually last longer because mechanical wear and heat stress are reduced.
Action: Turn off your computer at night!
Security benefit: When your turn your computer off, you decrease the risk of someone accessing your files or e-mail.

Myth #2: Computers don't really need a lot of power if they are on but not used—Wrong!
Fact: During heavy usage (e.g., when you open a new application), your computer draws only slightly more power. The average computer uses about 150 watts (75 watts for the screen and 75 watts for the CPU) whether you're using it or not.
Action: Turn off your computer if you are not using it for 1 hour or more!

Myth #3: Screen savers save.
Fact: Despite the name, screen savers do not save anything, especially not power!
Action: Turn off your monitor if you are not using your computer for more than 15 minutes!

requests for these materials, and we know that they have been distributed in Fortune 500 companies, as well as on campuses and other institutions on at least three continents. See figures 10.1 and 10.2. In researching the information for the computer-use brochure, special care was taken to contact computer manufacturers to obtain authoritative information on the question of whether frequent shutdowns are harmful to equipment. This inquiry revealed the unanimous view of computer manufacturers that shutdowns are positive.

At the same time that TCI prepared the brochures, we knew that simply reading a leaflet was unlikely to produce a great deal of behavior change. The field of social psychology has explored factors associated with behavior change in a range of environment-related behaviors. Much of this inquiry was initiated around the time of the energy crisis of the early 1970s. The field of social marketing builds on this knowledge, and later in this chapter we discuss a social marketing campaign to inspire students to turn computers off.

Climate change is real...

Turn off your computer!

Turn off your computer at night and when you are not using it for several hours.

Enable the Power Management feature for your monitor (see back).

Turn off your monitor when you are not using your computer for 15 minutes or longer.

If you buy a new computer, consider a laptop. Laptops use only 1/4 the energy of a desktop.

If you buy a new monitor, consider a flat screen. They use only 1/3 the energy.

Learn more, look inside!

Figure 10.1
TCI computer brochure

Figure 10.2
TCI computer sticker

Lighting and Behavior Change

TCI focuses on task and chandelier lighting as a way to increase awareness of energy efficiency and demonstrate alternative technology. We give away compact fluorescent lightbulbs, free for on-campus use. Whenever possible, we do "charge" for them—asking that people provide us with their old incandescent lightbulb. This program has distributed over 3,000 bulbs, including a retrofit of all lamps in the president's office. This lighting effort is part of a more comprehensive retrofit of the hardwired lights throughout the university.

A subset of lighting is the light in power projection machines. These projectors are increasingly common in classrooms, and they are very commonly left on. Educating room users about the need to turn these off can save energy, and more importantly, extend the life of the projector bulb. Since access to these projectors is difficult and lightbulbs are expensive, convenience should be an important part of a "power-off" campaign. In some circles, there is a myth that turning equipment off and on takes more power than leaving it on and shortens the lamp's life. While there is a momentary spike in electricity demand when the light is turned back on, the total electricity use (the area under the curve) is less for a given 24-hour period than if the lamp is left on during the night when the projector is not being used. Turning off a lamp over and over will increase the hours of useful life, since a lamp that is burning in an empty room will have no utility.

Vacation Shutdowns

Vacations, particularly those around the midsemester break, are times when dedicated efforts to shut things off and turn temperatures back can pay off. At Tufts the last day of exams for the fall semester in a recent year was December 20, and spring classes started on January 20. If students had early exams or papers, they were away for more than four weeks. The Tufts Climate Initiative and the energy manager developed a plan to target student rooms for an intense shutdown. Prior to our "shutdown project," Tufts facilities managers and the fire marshal's office conducted a visual inspection of windows (done from the outside of buildings) to identify and shut any open windows to prevent freeze-ups. The "shutdown project" involved several new elements with a goal of shutting down heat and equipment as much as possible so minimum energy would be used. Our effort included:

• Direct communication (postcards, e-mail, and poster) with students about what we wanted them to do, including turning down radiators, shutting and locking windows, turning off electronics, and cleaning and shutting off personal refrigerators.
• Training of resident directors.
• Visits to every room by a staff team, who confirmed that the effort had been accomplished and who fixed problems. (We found that twenty-six people in teams of two could visit about 2,000 rooms in a day and a half.)
• Setback of building temperatures.

The project has been conducted twice and is likely to continue. We have learned that direct and targeted communication with students can be effective, that students can and will take time to clean refrigerators (about 70 percent did so in the second year), and that the weather can make a difference, since moving out in an unseasonably warm week leads to many more open windows (as many as 18 percent in the first year when it was warm). The room visits had the added benefit of providing opportunities to discuss what equipment is in the rooms and what should and should not be allowed from an energy perspective, to identify and fix building problems, and to identify and fix safety problems. After the first year, we were able to combine this project with several energy-efficiency projects to increase heating control in residence hall rooms and to increase the spaces where we could set back temperatures. Knowing that the buildings did not have open windows or radiators calling for heat gave the facilities department greater confidence that temperature setbacks would not be problematic.

At Tufts we have also tried working on buildings that are shut down in the summer to unplug refrigerators and other appliances. To do this we have partnered with the residential facilities department and the custodial staff to create training about refrigerator cleanout, adding a "shutdown equals unplug, and leave the door open" component. An audit of some buildings during the summer or vacations will help to find the opportunities on any campus.

Operate Windows Thoughtfully

An open window is a wonderful addition to almost any room. However, in a conditioned room or building (those with heat or cooling), an open window can increase the demand for heat or air-conditioning. Building occupants need to be trained to shut windows when they do not need to be open and to understand how an open window affects others. For example, a window opened to cool an overheated office or classroom may eventually supply the nearby thermostat with that same cool air, calling for more heat and exacerbating the problem. Alternatively, buildings can be equipped with switches that turn HVAC off when windows are open, creating a feedback mechanism.

In our climate, the stakes are higher than saving energy dollars and reducing emissions. Right before classes started one January weekend evening, a dinner was held on the top floor of a multistory academic building. A window was left open in the kitchen. That night a cold front delivered brutal overnight temperatures, resulting in a single frozen pipe that fed a coffee maker. The frozen pipe burst. Consequences included a seven-story icicle that was about ten feet wide on the north side of the building, and water damage in faculty offices and classrooms on the second to sixth floors. These experiences, as well as others less dramatic on many campuses, show that window use must be responsible.

Personal Responsibility

There is an element of personal action for climate change and energy efficiency that includes personal responsibility and awareness. This is hard to teach (and perhaps harder to learn). In many ways the climate change problem and our inability to come to grips with it is a tragedy of the commons—a failure to consider how our individual actions have benefits or costs for the community. Personal responsibility for climate change action on campus might include:

- Tolerance for indoor temperatures that are cool in winter and warm in summer
- Dressing appropriately for the season
- Noticing and reporting problems with buildings
- Selecting energy-efficient equipment
- Biking or walking rather than driving
- Shutting off equipment when it is not in use

Social Marketing

Social marketing uses insights from social psychology and the techniques of commercial marketing to influence behaviors. As we have noted, environmentalists often assume that if people understand the problem, they will take appropriate action. Over the last twenty years, research has indicated that even when people understand the problem, they will not necessarily exhibit environmentally sound behaviors.[16] The discrepancy between environmental values and action is commonly referred to as the "value-action gap,"[17] and social marketing attempts to close this gap by using marketing techniques to modify behavioral norms in fields such as health, safety, and in this case environment.[18] Part of the strategy relates to the message itself. Telling teenagers that smoking was harmful did little to reduce their tobacco use. But in Massachusetts, when the message was changed to "smoking is not cool," smoking rates dropped among teenagers. The other part of the strategy requires going beyond messages to community norms. In other words, people need to be convinced that there is a group of people in the community (residence hall, department, club, or city or town) who find the information compelling and have changed their behavior in response.

Kristin Marcell, a graduate student in our Department of Urban and Environmental Policy and Planning, undertook a pilot-scale social marketing experiment in which the goal was to encourage students to turn off their computers when not in use. The experiment was conducted in two roughly comparable suite-style residences occupied primarily by juniors and seniors. In the control residence, a "knowledge-based" educational program on climate change was implemented. The program explained how electricity and consequently computer use generate carbon dioxide emissions. In the experimental residence, students received the same "knowledge-based" educational program as well as a social marketing campaign encouraging them to turn personal computers off when not in use. Before-and-after surveys were conducted, which suggested that the social marketing campaign had a greater impact on student environmental knowledge, attitudes, and behaviors than the "knowledge-based" educational program alone. However, the numbers of participants in both the treatment and control groups were small.[19]

One of the most striking aspects of Marcell's experiment is the materials she prepared for the social marketing campaign (see figure 10.3 for an example). This is a reminder that communication among students is likely to be very different from messages initiated by faculty or staff, and effective advocates will use the language of their audience. On campuses that offer degrees in communication, efforts to develop and test effective messages for different groups may form an important element of climate change action, and may represent significant contributions to the field.

We think this modest experiment is worth reproducing on a larger scale because the potential reductions in energy use are significant. The

Figure 10.3
Where's your hot spot?

Tufts Climate Initiative estimates that if only half of our residential students turn off their computers every night throughout the academic year, annual savings of approximately $30,000 could accrue. At a larger university, savings would be considerably greater.

This discussion of social marketing brings to mind the communication opportunities associated with emission reduction actions and interpretive signs. There are signs on the hybrid and electric vehicles, and faculty, students, and staff notice them and comment favorably. Experimental plots of native plantings have interpretive signs and also produce favorable comments.

One of our priorities is installing real-time energy monitoring and data displays in at least one student residence. Donella Meadows, whose work on systems thinking and commitment to the environment inspired a generation, described a subdivision of identical houses, but some had the electric meter in the basement and others had the meter in the front hall. In this example, which Meadows suggested might be apocryphal, electricity use by people with meters in the front hall was 30 percent lower.[20] That sounds like an experiment worth conducting. We think that providing real-time energy monitoring to students will provide multiple payoffs, creating opportunities for scholarly research as well as informing campus decision makers about the effectiveness of strategies for reduced energy consumption by individuals.

Inspiring Downstream Effects—Active Citizenship at Tufts

One of our goals in developing the personal action initiatives on campus is that all members of the university community will carry the lessons with them into other communities. For faculty and staff, we hope that learning about compact fluorescent bulbs and shutting down their computers at work will lead to greater learning and action that will in turn influence the decisions they make at home, as members of town government, through their activities in social service organizations, and in other aspects of their lives outside of work. For students, we hope these lessons will stay with them long after they graduate, and will permeate their professional practice and their home lives, including purchasing decisions and travel choices.

The Jonathan M. Tisch College of Citizenship and Public Service at Tufts is a nationally recognized leader in active citizenship. We actively partner with Tisch College because we see the personal actions being advanced by TCI as central to individuals' active citizenship, as well as to institutional citizenship and responsibility. As we undertake increasingly ambitious efforts to help members of the university community see the link between their decisions and impacts on the environment, we hope that the downstream effects of these actions will become increasingly significant. For example, we hope that members of the university community become active participants as well as advocates for building high-performance or LEED-certified buildings in their communities, and that energy efficiency becomes a central criterion in major purchases. We also hope that by having examples of successful energy-efficiency or alternative-energy projects on campus, members of the community will help increase the demand for new technology, solar, and other renewable-energy products through their personal and future professional purchasing actions.

The Debate: Pushing Personal Actions vs. Government Mandates

Kurt Teichert, Resource Efficiency Coordinator at Brown University, said, "Personal action is what you do while you are waiting for technology to be installed."[21] Clearly Teichert is a pragmatist who recognizes that even among those of us who are aware of our climate impacts, few are taking dramatic steps to reduce our personal contribution to them.

A question on which reasonable people can disagree is the extent to which personal actions to reduce greenhouse gases make sense. Ross Gelbspan, author of *The Heat Is On*[22] and *Boiling Point*,[23] contends that given the nature of climate change and the magnitude of the emission reductions that are needed, efforts to make changes should be focused on national governments. He argues, for example, that one of the most logical and simple ways to reduce emissions is for governments to mandate an ambitious fossil-fuel efficiency rate. We agree with Gelbspan that it is important to influence top government leaders. Global warming demands that national leaders act quickly and decisively to reduce heat-trapping gas emissions. While Gelbspan has cultivated access to leaders of many countries, in effect working at the top of the pyramid, the approach being taken by TCI works closer to the bottom of the pyramid.

We take the view that by working at the level of the academic institution and with the individuals in the institution, we can influence decisions in favor of emission reductions in the short term, although the reductions are necessarily modest. And we hope that with this approach, we can also influence decisions at the national level and over the longer term. We are particularly mindful of the influence of our graduates. Personal action may have a multiplier effect, as students become the policymakers, investment bankers, and business professionals of the future. For example, many Tufts graduates take positions in local, state, and national governments, often working in programs related to climate change and green buildings. These professionals understand the link between individual and organizational decisions and the problem of heat-trapping gas emissions. The challenge is to make that link for all members of the academic community and to address the urgency of the climate change problem.

A focus on personal action can have benefits for increased awareness and willingness to embrace new technology. Personal action can, in fact, make a difference, particularly when the rewards are personal, such as saving money at home. At Tufts, early efforts to shut off equipment in the dining halls resulted in one employee reporting significant savings at home by employing the same techniques there. This may sound modest, but marketing experts understand the value of positive buzz, and it is a great opportunity for climate action advocates.

There are, of course, other forms of personal action related to climate change. Direct political action including running for office, campaigning for legislation, and organizing protests are all options. If tens of thousands of college students were to take to the streets to protest government inaction on global warming, there is good reason to think they would be joined by tens of thousands of other concerned citizens. Considering that the effects of climate change will fall most harshly on people who are least affluent, least resilient, and least responsible for creating the problem in the first place, there is a great deal to protest.

Take the Initiative

How do you create a community in which people take personal actions to reduce their emissions? Perhaps it begins by creating an expectation of efficiency and action. This expectation can be communicated repeatedly and used as a framework for requesting specific actions. We believe that climate action often lacks a coherent message, thereby creating the perception of a series of disconnected projects rather than a carefully orchestrated movement or campaign.

At Tufts, we are learning how to create a community of active citizens engaged in meeting this expectation. Our most resource-intensive efforts have focused on students and on select staff. We hope that this engagement will be the basis for inquiry and learning among climate advocates.

Involve Students Directly

One group whose personal actions deserve special attention is residential students. As we mentioned earlier, plug loads in campus residences are a source of increasing emissions, and as energy costs rise, an increasing financial burden to the institution. Establishing a strategy for plug loads in residence halls makes sense, although strategies will vary across institutions. Below are several suggestions for student involvement.

Start an Eco-Reps Program Modify community norms through peer-to-peer programs in conjunction with programs that influence purchasing decisions prior to students' arrival on campus. At Tufts, Eco-Reps are undergraduate students who commit to undertaking environmental actions on campus with an emphasis on engaging their peers and taking actions in their residences. The goal is to connect personal action and impacts with knowledge of the environment and engage students in campus environmental activities. The program helps increase overall student awareness of environmental actions and identifies ways to effect individual change on campus.

The Eco-Reps program is modeled after a Dartmouth program, and is a hybrid between a regular course and an internship. Students who are Eco-Reps have a noncredit class every other week and participate in several field trips. Each class is organized around a particular topic, including recycling and waste prevention, climate change, water resources, food and the environment, and consumption. Box 10.3 outlines a sample Eco-Reps program. There are seven 2-hour meetings throughout the semester. Eco-Reps engage in on-campus greening activities. Typical activities include checking on the recycling program (e.g., bin placement, contamination, and so on), talking to students, conducting surveys, putting up posters, and organizing awareness-raising events.

Box 10.3
Sample Eco-Rep activities

Activities
During each meeting, students are given a project sheet with a theme that specifies what the requirements are (including estimated time for each activity). Here is a sample of a typical two-week activity sheet:

• *Read a chapter in the Eco-Rep Manual (15 minutes).* Eco-Reps were given a manual at the beginning of the semester to use as a guide and resource for further information on the topics covered each week.
• *Interview students in your dorm (40 minutes to 1 hour and 20 minutes).* TCI staff developed a questionnaire for students asking about their level of awareness of environmental issues, individually, on campus, and globally. The goal is to educate the Eco-Reps while conducting the questionnaire (with five students) and to start discussions about some of the topics the program will address during the semester.
• *Carry out a Climate Change Education Action (1–2 hours).* Guided by three goals—educate people about climate change, educate people about TCI, and get people to conserve energy—Eco-Reps were asked to create an action, define the message, identify barriers to action, and develop a plan to address the barriers. Students also evaluated their activity before and after the action to understand the success.
• *Continue to check recycling in the dorm (30 minutes).* In one of the first meetings, students were asked to evaluate and monitor recycling in their dorm. Each week, part of the Eco-Reps' responsibility will involve checking to make sure that recycling is working (if not, identifying and solving problems such as not enough bins, unclear identification, contamination, and so on).

Focus
During the first semester the program focused on educating the Eco-Reps and organizing activities in their dormitories. During the second semester each Eco-Rep chooses one project and works on it in a small group for the whole semester. At the end of the semester, each group will give a 10–20 minute presentation about the project.

The following is a list of projects that groups have worked on in the past:

1. *Green Purchasing* This group worked with Tufts Purchasing on introducing 100 percent postconsumer recycled paper at Tufts. (We currently use this paper at TCI.)
2. *Environmental Writers* This group focused on writing articles for Tufts and local publications on a variety of environmental issues.
3. *Food Campaign* This group focused on educating students about the impact of food and food production on the environment.
4. *Individual Projects*:
• Improving recycling in the frat and sorority houses and dining

The program is a combination of information sharing—in the meetings, as part of the activities, and in written materials we prepare and distribute to the students. TCI staffer Anja Kollmuss strives to give students enough guidance so that they are not left hanging, at the same time providing as much freedom and support as possible.

Students receive compensation for participating in Eco-Reps; however, the $150 we are able to pay is modest at best. Students say that the money is valuable in keeping them going in a crunch. Although students are offered the option of having the program form the nucleus of an internship (required for undergraduate environmental studies majors), or for credit as an independent study project, thus far none have exercised these options. The Eco-Reps program does a good job of educating passionate self-selected students. Many join the program because they want to learn to distinguish the actions that make a large difference in environmental outcomes from the actions with minimal effect.

The Eco-Reps program on our campus is broader than climate change and includes the full range of campus greening issues. This means that students engage with recycling, purchasing, and infrastructure issues. For example, students conduct surveys on water leaks and heating to identify whether temperatures in dormitories and classrooms are comfortable.

Building Shutdown Programs When students leave campus for vacations, the college saves on energy bills if appliances are unplugged, windows closed and locked, and in the winter, radiators are turned down. Earlier we described a vacation shutdown effort designed to create a sense of individual responsibility. The success of these efforts requires communicating effectively with students, along with efforts to modify community norms.

Waste Reduction and Recycling As we discussed earlier, waste reduction and recycling have climate change benefits, so the university can encourage recycling and make it easy. Through a professional recycling coordinator, an Eco-Reps program, or another established campus activity, recycling in student residences can be a useful way to reduce the amount of waste, reduce emissions, and create an opportunity to convey other messages related to campus sustainability and citizenship.

Contests For several years, students at Tufts have conducted an energy conservation contest called "Do It in the Dark." The idea is that residence halls compete against one another, and the winner is the hall that uses the least electricity.

On our campus, it is extremely difficult—given the challenges associated with metering and establishing a baseline—to have a scrupulously accurate and fair determination of which group "won" when comparing energy use of student residential facilities. Students use a surrogate measure based on the previous year's consumption to determine a winner. Technical limitations aside, we think the program is terrific. We regard the contest as a valuable student awareness-raising activity, and support it when asked, but do not get involved in the details of how students make decisions to designate winners.

This program is conducted by the undergraduate environmental group known as the Environmental Consciousness Organization or ECO. Students in ECO select a time period (usually about four weeks) when the contest will occur, publicize the event, collect energy-use data, and designate the winning residence. The contest is extremely popular, in part no doubt because the name of the contest is appealing. The merchandising opportunities have not been overlooked by enterprising students and one year the contest featured boxer shorts with the "Do It in the Dark" logo, along with the perennial favorite, glow-in-the-dark condoms. When the contest is won by a hall for first-year students, the hope is that some consideration of electricity use will carry through the students' entire undergraduate career. As with many programs that rely exclusively on undergraduates, the follow-through can be uneven. Students vary in their organizational skills and interests, and quite understandably, their priorities change from year to year.

"Do It in the Dark" is an excellent example of a program that is well worth supporting when students provide the labor and the enthusiasm. It would be hard to argue that the program produces long-lasting emission reductions; however, that is not its primary goal. It is an awareness-raising campaign, and in an ideal world, participants will be more attentive to future messages related to energy use, personal action, and climate change.

Communicate with Both Parents and Students at Home It is common practice for colleges and universities to send materials to families detailing what students will need to have (pillows, sheets, blankets, wastebasket), what they may wish to have (study lamp, TV/VCR/DVD), and what they may not have in residence halls (toasters ovens, space heaters, candles, and so on). It is ideal to use this communication to encourage students and their families to bring only the most energy-efficient devices available to campus and to convey expectations. We find that sometimes students call home when they are not satisfied with the answer they get on campus, and a clear set of expectations from the outset will help to alleviate this problem.

- Send home mailings about Energy Star computers and other appliances and explain the campus commitment to reducing the effects of climate change. The assumption is that this material will be read by many parents, and may be most effective in influencing purchasing decisions of incoming first-year students. There also may be collateral benefits not captured by the college or university in raising parents' awareness about the opportunities for energy efficiency at home.
- Send home brochures for computer-off and lightbulb-exchange programs. We make this suggestion because we want both students and parents to understand that there are environmental and financial implications of everyday-use decisions. If both families and the college begin to deliver consistent messages, perhaps we will see student choices change in favor of reduced emissions. For example, as we saw earlier, if students elect to bring a laptop instead of a desktop computer, the student's energy consumption will be lower.

Link to Student Fees In response to unanticipated spikes in heating and electricity, some colleges and universities have added an energy surcharge[24] to students' bills, along with explanatory material.[25] The positive aspect of such an action is that it makes a clear link between the cost of residing at an academic institution and energy costs. However, there are also negatives. The arrival of an unexpectedly large bill may not be a particularly "teachable moment" for students or the person writing the check.

Encourage Faculty and Staff to Take Personal Actions

Faculty and staff are important to target in efforts to encourage personal action. As with students, these efforts must be ongoing and direct. There are many ways to build programs for faculty and staff that include written material, e-mail reminders, and demonstration projects.

Energy as a Priority Staff with direct responsibility for energy-intensive equipment use, such as staff in dining services, printing services, or facilities, should have energy use identified as a group and personal priority. By creating a "spot" for energy on department agendas, staff trainings, and other aspects of community life, staff will begin to identify significant areas where savings may result.

TCI has found many opportunities for energy-reducing partnerships with departments across campus. Some are lasting, while others entail periodic training or awareness efforts. In the mid-1990s we worked comprehensively with dining staff and have revived that effort recently. We also conduct regular workshops with staff in academic computing.

Demonstration Projects Brochures for computer-off and lightbulb-exchange programs can be followed up with additional action that will have meaning for the community. For example, it may be strategic to conduct an energy audit and lighting retrofit in a highly visible building on campus such as the president's house or the main administration building. When emission reducing actions are taken in buildings that are considered showpieces on campus, there is the added value of communicating to faculty and staff that energy efficiency is not being practiced at the expense of style and aesthetics. Campus news coverage and/or interpretive signs can call attention to both the activity and the technology used. We have found that even some committed environmentalists on campus were unaware that chandeliers can be fitted with small "flame-shaped" compact fluorescent bulbs. If a local vendor can be convinced to donate bulbs and lighting controls on a demonstration basis, it may be an opportunity to convey the message to faculty and staff that the materials to undertake retrofits at home are available in the community. The energy-saving workshops for faculty and staff mentioned earlier are another important opportunity to strengthen the link between actions at home and at work that can reduce emissions.

Purchasing Decisions Making available information on the effect of purchasing decisions on climate change for faculty and staff can be an important strategy (see chapter 8). Particularly if purchasing is decentralized, creative approaches to get the message across should be developed and implemented. At a minimum, an informational campaign should introduce the Energy Star website and explain the link between purchasing decisions and greenhouse gas emissions. Because many people associate recycling with personal environmental action, making the link between climate action and recycling can be useful. Staff and faculty need recycling programs that are convenient and visible.

Link Personal Heating and Cooling to Emissions Policies such as heating and space heater strategies (see chapter 9) are important to tackle in order to demonstrate institutional leadership that can be replicated by individuals. Perhaps this is a more salient issue in the Northeast than in other parts of the country; however, the link between personal heating and cooling and emissions reduction is an important one.

Link Energy, Budget, and Emissions Climate action programs can implement broad-based information campaigns: one person can take a bite out of energy consumption and climate change. On campuses experiencing dramatic increases in energy bills, the cumulative effect of individual actions really can make a difference. A broad campaign can have messages related to both home and work and can include highlights of lighting, computer use, purchasing, and other programs with references to a website and/or people for additional information. Encourage reporting of energy-related problems on campus (e.g., overheating or overcooling). This can be done by distributing information on where to call for different problems, and also can be supported by a building curators program, described in chapter 7.

Summary

Developing and implementing effective programs related to personal climate actions may be challenging but worthwhile. A focus on personal action allows us to expand the educational reach of our commitment to reducing heat-trapping gases. Even if the emission reductions are modest, personal climate action can be an excellent complement to other programs on campus that relate to active citizenship, social responsibility, and service learning. In addition, personal action may have downstream effects on decisions off campus or effects that manifest in the long term. In the next chapter we discuss climate action in the formal learning agenda.

11

Climate Action in the Classroom

In this chapter we focus on education for climate action using the campus as a learning laboratory. We find students attracted by the pragmatic nature of action projects, excited to learn about the impacts of apparently mundane decisions, and eager to generate information related to the implementation of international climate change policy. Climate action projects can help students in a wide range of disciplines think about the implications of their decisions for the environment, and can foster learning in a wide range of traditional disciples and a host of interdisciplinary inquiries, including energy and energy systems. Students bring enthusiasm and new ideas to the campus, and staff, particularly in the operations division, bring knowledge, pragmatism, and experience to student learning.

One of the greatest challenges in taking climate change action projects into the classroom is learning how to manage the effort so that it is rewarding for students, meets the learning objectives established by the instructor, and produces useful and timely outcomes for decision makers. Certainly not all climate change action projects have the same challenges, but here we discuss some that have emerged repeatedly from our experience.

Climate Change and Academics

Climate change can play a role in virtually any classroom. The intergenerational impacts of climate change invite discussions from philosophers and poets. The technology to mitigate emissions of heat-trapping gas calls on the work of a wide range of engineering and scientific disciplines. Political scientists, economists, policy developers, planners, and diplomats have roles in examining the short- and long-term implications. Biologists, epidemiologists, and geologists will continue to generate data to help us understand the rate of change and the impact on natural systems. In addition, climate change should be considered a predictor of change and a causal agent for biological, political, and meteorological events.

These are some of the more obvious relationships. Less obvious, but equally important, are the impacts on business and on the financial health of a wide range of institutions. The insurance industry already is profoundly affected by climate change, and indeed, was one of the first business sectors to "discover" the phenomenon. Energy companies have followed, but many more are just waking up to the long-term financial implications of climate change. It follows naturally that business educators will have to increasingly help students understand and take into account climate change as they consider the viability of investment decisions.

Art offers a tremendous opportunity for climate action advocates. We have seen the power of public art in motivating and educating about a range of social issues. Artists may be able to inspire widespread climate action far more effectively than scientists and planners. A friend suggests that a high-tide line be painted on all the buildings in Boston to show the effect of a specific amount of sea-level rise. With too much sea-level rise, Boston will of course be inundated. If we unleash the creativity of climate-knowledgeable artists, what powerful images will they produce?

Develop Faculty Interest in Climate Change Education

Climate change can be explored through many disciplinary lenses and it cuts across a variety of disciplines, placing many salient questions squarely in the realm of interdisciplinary studies. However, it is likely that faculty in some disciplines may not see the connection between their teaching and research and climate change. Creating learning opportunities for faculty can take several forms, ranging from bringing engaging speakers on campus to address the issue from a wide range of disciplinary and interdisciplinary perspectives, to forming study groups, to launching a climate change development program.

In 1990, Tufts pioneered a faculty development program called the Tufts Environmental Literacy Institute (TELI), featuring an intensive immersion in environmental issues across disciplines. Emory University now has a similar faculty development program on sustainability called the Piedmont Project.[1] We think a comparable effort is needed for faculty on the issue of climate change. As with TELI, an effective outcome would be faculty commitment to curriculum changes that link existing material and climate change, or add climate-related content to student projects or academic research.

Each module of a faculty development program can address the elements of climate change education described in the next section. The goal is to develop a general understanding of the issue and its complexity, and at the same time encourage participants to consider what they can do to incorporate climate change in their teaching and advance climate change action starting tomorrow. The material is general so that it can be used in a variety of learning environments but can also be modified for use in a particular disciplinary context. The assumption is that a climate change development program may have the result of modifying faculty decisions at home and at work, as well as motivating creative opportunities for introducing climate change into the curricula for courses in many disciplines.

A faculty development program that focuses on climate change is ideal for an enterprising provost to support, and the program can be implemented either internally or in partnership with other colleges and universities.

Elements of an Action Program for Faculty

Understanding the climate change problem does not automatically inform solutions, and it is important to make this distinction when developing a strategy for taking climate change into the classroom. And we know that framing matters: climate change is an enormous societal challenge and it should not be marginalized as an environmental problem. The following include some aspects of climate change action that can be incorporated in a faculty development program and in whole or in part into the college or university curriculum.

What Is Climate Change and Why Does It Matter? Offer a brief historical perspective on climate change science and policy, build the case for climate change action in the classroom, and end with an exercise in which participants draft a preliminary action list that is relevant to the participants' academic expertise:

- Climate change science, climate change policy, and climate action
- Implications of a changing climate for relations among countries and for people already in poverty
- Actions that can be taken

How Do Energy Decisions Affect Climate Change? Raise awareness of climate change issues and offer a framework for considering changes in practice:

- Energy and the climate—a basic overview of air, water, and waste issues associated with the extraction and use of different energy sources (including coal, oil, natural gas, and biofuels) and different electricity-generating technologies
- Energy reliability and cost in a changing world
- A framework for assessing energy sustainability
- Energy management on campus—responsibility and authority; implications of incentives and disincentives
- Assessing our institution's energy strategy
- Energy strategy in our region and at the national level—implications of incentives and disincentives
- Comparative national strategies

What Is the Link between Water and Climate Change? Raise awareness of the links between water quality, quantity, distribution, and use on the one hand and climate change on the other. Select examples to include intensive discussion of sea-level changes if appropriate (these will include saltwater intrusion on freshwater supplies as well as a wide range of impacts on water and wastewater infrastructure in coastal communities):

- Overview of human impacts on water quality and exploration of water-use patterns
- Linking water and climate change
- Role of governments in quality, pricing, treatment, delivery, and managing supply
- How much does water cost and how much do you pay?
- What goes in must come out—thinking about wastewater as a resource

- How climate change affects water supplies
- A framework for thinking about water use in the institution

What Are the Economic Implications? Begin with an examination of the economic debates surrounding the Kyoto Protocol and explore alternative approaches to achieving emission reductions—for instance, by using a per capita approach. Explore the very different equity effects of these approaches along with the efficiency arguments. Topics to consider:

- Climate change and development
- Markets for emissions
- Technology policy, innovation, and climate change
- Domestic policy considerations—for example, how much does transport of goods by truck cost and how much does industry pay? How much does automobile travel cost and how much does the driver pay?
- International law and diplomacy

What Are the Policy and Planning Implications of Climate Change? Planning considerations can include mitigation as well as adaptation measures:

- Planning for natural disasters, especially those whose frequency is likely to increase as a consequence of climate change
- Master plans, emergency plans, financial plans
- Creating communities that emphasize pedestrian access and transit use
- Planning for communities that will experience sea-level rise—the issue of managed retreat
- Policies and plans that emphasize efficient infrastructure that minimizes reliance on fossil fuels
- Planning for energy supplies that are reliable and not subject to price volatility
- Policies that discourage climate-sensitive change

What Can One Individual Do? Making explicit the link between emission reduction practices at home, on campus, and in the workplace can help to foster development of creative solutions in the classroom and in the larger campus community. This discussion is suitable for a wide range of students, faculty, and other employees, and focuses on raising awareness and teaching simple steps that can be taken at home to reduce emissions:

- Help participants understand how to reduce energy and water consumption at home (and/or other goals suitable to the specific geographic context in which participants are located) and link these actions to climate change
- Discuss technology options (dual-flush toilets, insulation, lighting upgrades, solar thermal, photovoltaic, computer settings, computer choices, and so on) as well as behavioral changes
- Raise awareness of issues associated with the power of household decisions, including purchasing locally produced goods, purchasing green power, supporting family-owned businesses, supporting people with disabilities, supporting minorities, and so forth
- Raise awareness of the climate change implications of personal transport choices
- Discuss the role of climate change and political choices at the national, state, and local level, as well as the role of voting
- Identify steps that can be taken at home tomorrow

Climate Change Education within Academic Disciplines

For courses taught by a single faculty member, it may be quite straightforward to introduce climate change in the classroom with the goal of raising student awareness. Box 11.1 has an assignment that can be adopted for several disciplines. We use it in a course on climate change. But there are more subtle ways to introduce global warming–related material. For example, in a quantitative methods class, a decision can be made to have students learn methods using data sets related to climate change. Some of these data may be meteorological, and accessed through a wide range of Internet sources, including NOAA and the World Meteorological Organization, while others may be specific to campus.

Using campus data, students can examine relationships between ambient temperature and fuel use, and can examine electricity use before and after lighting upgrades in individual buildings. When a college or university has historical data, students can learn which segments of campus energy use are increasing over time and which are remaining level or decreasing. For instance, students may find that fuel for heating is decreasing per square foot, but that electricity use is increasing per student. Engineering students can assess alternative approaches to

Box 11.1
An awareness-raising assignment

In a policy and planning course on climate change, you can send students to the newspaper. This assignment can be modified to fit several disciplines.

In *Boiling Point*, Ross Gelbspan, a former *Boston Globe* reporter, is very critical of media coverage of climate change. This is an issue we will examine in some detail as we explore public understanding and later take up specific issues such as energy and transportation systems. For now, we will scrutinize newspaper coverage for climate change, examining opportunities taken and not taken. Please examine the *New York Times* for [a specific date].

1. List the articles in the paper that mention climate change (if any). For each, indicate the following:

- Headline of the article
- Position in the paper (page number and section)
- Theme of the article; select one (or more) of the following themes for each article:
  - International relations
  - Domestic politics
  - Consequences of climate change
  - General science
  - New evidence
  - Controversy among scientists
  - Economics
  - Alternative energy
  - Other (you specify)

2. List the articles that could be related to climate change, but where the relationship is not established in the article. Find as many as possible and do be imaginative. For example, an article might address fuel-efficiency standards for cars, or home insulation, but might not mention a relationship between reducing energy use and reducing heat-trapping gas emissions. For each, indicate the following:

- Headline of the article
- Position in the paper (page number and section)
- Theme of the article
- The link with climate change that the article did not make (explain in a few sentences)

Please do not limit yourself to news articles. Material appearing in any section of the paper (even the sports section and obituaries) is fair game. After you have completed your analysis, conclude with a paragraph or two about what you found and what effects you think this news coverage will have on public understanding of climate change. Three pages total.

heating and cooling particular buildings, and a social psychology class can learn about social marketing using climate change action as the social goal. Table 11.1 gives examples of topics within disciplines that can be enhanced with global warming content. For activities of this type, the primary challenges include ensuring that the course instructor is well versed in climate change issues, and making sure he or she plans far enough ahead to secure access to data that suits the needs of the course. If a goal is to use campus data, faculty will have to ensure that data sets exist in manageable form and can be made available to students.

Climate Change as Interdisciplinary Study

Having mentioned several traditional academic disciplines, it is important to note that most of the climate change efforts at Tufts are interdisciplinary at some level. Although Tufts values and nurtures interdisciplinary work, it has to be acknowledged that a variety of challenges remain. Some courses with climate change content are cross-listed in more than one school or department. One course is cross-listed in three schools (Engineering, Arts and Sciences, and Law and Diplomacy). This offers a fabulous opportunity for very different kinds of students to meet one another and grapple with the same material from a variety of perspectives. At the same time, students and faculty can face challenges with these arrangements related to scheduling, calendars, background knowledge, and expectations.

For courses that are cotaught by faculty in different disciplines, discussions with department chairs may be needed to emphasize the value of the course to each department. A climate change development initiative for faculty can have enormous value in acting as a catalyst for interdisciplinary collaboration on campus. If the development initiative is valued by top decision makers, it may help break down institutional barriers to interdisciplinary teaching and research.

Some climate-related courses or course modules that lend themselves particularly well to interdisciplinary approaches include systems analysis and system dynamics, scenario projections of climate, economics and climate change, impacts on ecosystems and the economy, modeling of complex systems, complex multiparty negotiations, and ethics and values.

Table 11.1
Making the climate connection within disciplines

Discipline	Examples
Political science	Subnational initiatives for emission reduction, positions of political parties, grassroots actions, vested interests
Economics	Costs and benefits of emission reduction alternatives for campus action, assessment of competing economic models of climate change and their underlying assumptions
Policy and planning	Community resilience, future-growth planning informed by adaptation, precautionary principle, natural-disaster response planning, infrastructure planning
International relations	Equity assessment of Kyoto Protocol and alternative international regimes, UN, IPCC
Engineering	Data collection and analysis, development of renewable energy, application of climate models to regions and subregions, assessment of climate change implications for regional infrastructure, life-cycle assessment of biofuels such as ethanol and biodiesel
Anthropology	How people understand climate change; variations in understanding across national boundaries
Public health	Response to extreme weather events, water supply protection
Sociology	Culture of consumption; social movements
Literature	Future scenarios, nonfiction accounts of people and places
Mathematics	Analysis of data sets related to climate change (e.g., atmospheric carbon dioxide, emissions, weather patterns.)
Geology	Melting glaciers and permafrost, ice-albedo feedback
Art	Public art projects designed to inspire inquiry and action, including photographic images of extreme weather
Philosophy	Intergenerational issues, equity issues, moral implications of our current actions, individual and group responsibility
Biology	Migration of plant, animal, and insect species, carbon dioxide sequestration
Epidemiology	Changing patterns of tropical diseases, West Nile virus, eastern equine encephalitis

The Intersection of Academics and Operations

One of the great unexplored opportunities for synergy in the university community may be faculty and their students working with administrators and staff to address climate action on campus. There can be a chasm between faculty and student activities and those of the operations staff and top decision makers, but it can be bridged. There are institutions where this problem is virtually nonexistent, so we know there is wide variation in practice.

Too often, faculty on our campus have expressed interest in campus projects, especially new buildings, when it is "too late" in the project planning cycle to incorporate ideas without significant cost penalties. This problem can be overcome with a more transparent process for decisions about new facilities. Creating a culture in which faculty consult with campus decision makers as they plan courses in much the same way they research the literature and examine the latest texts in the field greatly enhances the opportunities for projects that can have a positive influence on campus decision making. Operations personnel have significant incentives to collaborate with faculty, because they can advance their knowledge with respect to emerging ideas and technologies and become integrated in the core mission of the university: educating students. Faculty have significant incentives to collaborate with operations personnel, so that they and their students develop an understanding of the practical issues associated with identifying, selecting and implementing, monitoring, and evaluating new or modified technologies and systems on the campus.

Another strategy is for faculty to serve on campus committees that can influence climate change action. Our belief is that the more engaged faculty become in planning and developing the college or university infrastructure, the more opportunities they will have for interacting with administrators and operations staff. In some cases, these interactions can be transformed into opportunities to take climate change action into the classroom and then toward full-scale implementation.

Lastly, faculty and students interested in climate action must realize that conducting a study or suggesting an idea is not necessarily sufficient to motivate change on campus. We have observed faculty who are content to offer solutions and then walk away, perhaps assuming that their suggestions will be followed automatically. One person attended a single master plan meeting, presented his idea, and then declared, "I have done all I can." The theory that knowledge is sufficient to motivate action has been disproved repeatedly in our work. In many campus cultures new ideas, whether they involve technology or strategy, must be comprehensively worked into the decision-making fabric of a project plan and funding scheme, often balanced against competing priorities. In most cases, a great deal more effort must be expended to move from ideas to action, and that is the effort that requires so much TCI time.

Preparing for Action in the Classroom

There are many opportunities to use climate change in the classroom, both to enhance student learning and to inform climate change decisions. However, substantive challenges often emerge if the central goal is to produce information useful to the campus agenda for climate change action. Some of the challenges are associated with problem-based learning in general and some are particular to the types of decision makers on campus and the information that needs to be generated. The Tufts Climate Initiative and other sustainability efforts have had over 250 student projects focusing on campus-related issues, and while we have made many mistakes, we have learned in the process. Below are some reflections on successful and unsuccessful student projects designed to educate students and inform climate change decisions on campus.

Because Tufts has made a commitment to meet or beat the emission reductions associated with the Kyoto Protocol, this goal represents our point of departure for student projects. Our greenhouse gas inventory was refined by a postdoctoral student, but in other organizations, it may be easier to assemble a good inventory with staff or faculty expertise. The inventory is a very important step, and the postdoc who did the work for TCI was exceptionally well suited to the task by virtue of both academic training and personal attributes.

Institutions may initiate climate change efforts by asking students to examine the implications of a similar or more ambitious emission reduction goal and to begin constructing an inventory of heat-trapping gas emissions. Because we have a clear goal, many aspects of our planning are facilitated; however, the goal has a direct impact on the type of student projects we use our time to support. We constantly ask ourselves whether the outcome of the project will result in reduced emissions from university operations. If the answer is no, we ask ourselves whether the increase in awareness that might result from the project justifies our involvement. Or we consider the value of projects related to adaptation, knowing that some adaptation to a changed climate will be needed. Often we find ourselves struggling over potential projects that are absolutely fascinating but bear no direct relationship to the task at hand. We engage in some, knowing full well that we are sacrificing emission reduction opportunities.

This discussion will begin with some general observations about taking climate action into the classroom and then will follow with specific suggestions for undertaking such efforts. Again, the primary emphasis is on projects that will lead to increased student learning and information that will help reduce emissions.

Characteristics of Successful Projects

Successful student projects related to climate change that inform decisions and action share several characteristics.

Projects Have an Identifiable Client These projects have an identifiable "client" who is able to answer student questions and who conveys a genuine interest in the outcome of the student work. A client is a person who expects to use the information students generate. Clients for student projects have included TCI staff, the energy manager, the director of dining services, the purchasing department, the public safety department, and members of the facilities or construction departments. Sometimes TCI is the client, acting on behalf of a group in operations, and sometimes the client is the individual decision maker such as the energy manager. If background information or historical data are needed for students to perform their work, these will be made available, or alternatively, students should know in advance that a significant portion of the project will be the collection, consolidation, estimation, and/or organization of these data. It is critical that students or faculty initiating projects recognize the value of and the demands on operations staff for these projects. As the client, staff should be treated as the customer rather than as a last-minute resource.

Select Good Clients What makes a good client? Good clients can emerge from any level of the organization and share key characteristics. They

- Make time at the beginning of the project to meet with students to provide background information and discuss the problem statement and their expectations.
- Negotiate the scope of work as necessary to reflect student strengths and weaknesses.
- Prepare in advance any materials that students will need, and, if applicable, advise others in the organization that a student team is working on a project and may be requesting information.
- Do not change the problem statement or scope of work halfway through the semester without first consulting with the faculty member in charge of the project and then the student team. If a significant internal or external event occurs that "should" be factored into the project, the faculty member may decide that given the learning objectives of the course and the strengths of the students, it may make more sense to allow students to continue with the original plan. If this is the case, the value of the project may be greatly diminished from the client's perspective, but a good client understands that when a potential conflict arises, faculty have to decide in favor of a positive learning experience for students.
- Understand that students may be caught in the crossfire of competing expectations. For example, the client may tell students that he or she has no interest in a bibliography and is virtually certain not to read it. The faculty member may tell students that a good bibliography is a requirement of a good project. A good client, on learning of this inconsistency, will remind students that a project of this type is somewhere between a real work experience and an academic one, and they need a good bibliography to meet course requirements.
- Make time during the project cycle to read a draft or drafts and provide comments if requested by students.
- If there is a student presentation, attend the presentation and ask challenging questions.
- Think carefully about how the project will be used and communicate it clearly to students. If the project is intended primarily to explore an option with the likelihood of dismissing it from future consideration, students will not necessarily be less motivated, but they may craft their work differently than they would knowing that their recommendations will influence decisions taken within a few months.

- Manage their own expectations about the value of the project in informing decisions, but remain open to the possibility that they will receive an extraordinary product.

Manage Clients Just as there can be student teams that disappoint, there can be clients that make students' lives miserable. Our experience suggests it is highly desirable for faculty to interact with prospective clients the semester before the project is launched. Strategies that can be followed by faculty to minimize problems with clients include:

- Meet with prospective clients or talk with them on the telephone about expectations.
- Have a sufficiently detailed understanding of the client's proposed project so that the faculty member can prepare an abstract or preliminary problem statement, and is satisfied that there is a good match between client needs and learning objectives for students.
- Provide clients with a written list of expectations.
- Take note of particularly good or particularly poor clients, and convey this information to others on campus who engage in climate action projects.

At the same time that faculty must manage clients, they also have to manage students. We all know students who produce outstanding term papers at the absolute last minute, usually the weekend before the due date. Deadline-defying virtuoso performances can be impressive, but they are fairly certain to cause meltdowns if done as part of a team effort, and can annoy a client who expects drafts on an agreed schedule. Here are some approaches to managing students and their projects.

Projects Begin with a Problem Statement In general, a successful project either begins with or asks the students to generate a succinct problem statement. It is critical that the problem statement be very carefully designed to reflect a problem that is manageable, for which data are to be collected or already exist, is amenable to analysis by the students (not outside of their technical expertise), and is something that the client can assist with. Appendix E shows examples of how TCI crafts these problem statements in the form of project descriptions. Often posed as a question, the problem statements may be framed by TCI and subsequently modified by students. For example, we are interested in identifying furnishings and office products that will reduce emissions of greenhouse gas. This interest was a good fit with an undergraduate environmental economics course in which the professor wanted students to conduct cost-benefit analyses. TCI asked student teams to focus their cost-benefit analyses on a single category such as floor coverings, computers (desktop versus laptop), and integrated office machines (fax and copier together versus separate).

One strategy for making student projects an effective part of decision making is for faculty who wish to sponsor projects to ask decision makers what would be useful to them. Too often, this fairly obvious step is overlooked. This can lead to faculty designing projects that are well intentioned but not particularly useful. This is a poor use of everyone's time, especially when a modest shift in project focus might be more valuable to decision makers.

Often refining the problem statement requires extensive interaction between students and the project client. It is also possible that the problem statement should be the subject of preliminary negotiations between the faculty member teaching the course and the client. This is particularly the case if the client is not able to articulate a problem statement that students will readily understand, or if the problem statement is wildly unrealistic given time and student skills. The faculty member also has a responsibility to ensure that the academic objectives of the project are met, and this can often be communicated most effectively through a discussion with the client and/or a written summary of expectations provided to the client before the project begins.

Students as Data Collectors In general, projects that place emphasis on gathering published information are good candidates for student projects. Advocates for climate action often lack time to conduct background research, so this student strength plays to a real need. With minimal guidance from a resource librarian, students can identify and secure access to a wide range of published data.

Although it may require considerable effort on the part of the instructor or client, students also can be an excellent resource for generating new data. Depending on the nature of the course, generating new data may be a central focus of a collective effort or it may be a relatively small element of an individual or team effort. For example, in a course on survey design, a political science faculty member had the class develop and field a survey of undergraduates. TCI worked with the faculty member to ensure that several questions related to climate change knowledge and action were included in the instrument. Part of the students' learning related to sampling strategies and data integrity, so the course instructor retained responsibility for this critical aspect of quality control. On the other hand, when data collection is a small element of the course's learning objectives, responsibility for quality control may shift to the decision makers who seek to use the student-generated information in decision making. For instance, in a course on environmental technology, a small team of students conducted an inventory of boiler makes, models, and rated efficiency in the small wood-frame buildings on campus. In this case, it was in the interest of TCI to work closely with students to establish a data reporting and quality control strategy so that decision makers in the operations department could be confident in using the data.

Although it is widely understood that collecting data is expensive and it is assumed that engaging students in the process will greatly decrease costs, this is not always the case. Depending on the circumstances, it may require significant resources to train students to collect high-quality data. Without extensive involvement of the decision makers or their proxies (such as TCI), there also is the risk that data collected are not useful in revealing information about the phenomenon of interest.

Care must be taken to direct students to data that can be gathered, rather than to data that we would like but have no way to gather. One example is that each year several students ask us for data on steam use by each building. While we too would like these data, they are not currently available because we do not have individual steam meters and because measuring low-pressure steam is challenging.

Student data collection efforts (and all data collection efforts) should be guided by the research question, rather than by mere curiosity. For example, we are often asked "How much electricity does Tufts use?", but rarely does that information inform any inquiry that is being undertaken. On the other hand, students have been successful in uncovering useful information that people have been reluctant to give directly to TCI. For instance, when we were exploring possible applications for our first hybrid vehicle, campus police personnel told students they thought the vehicle was too small, an opinion they did not disclose to TCI. Similarly, students have been very successful at learning about climate actions at peer institutions. We learned a great deal about energy reduction in tel-data closets from a student who had a friend in the right place at another university, and are now pursuing changes that will have short- and long-term payoffs.

It is hard to put too much emphasis on the quality control aspect of student projects. In one of our favorite projects, the student began by asking "How much does a gallon of water cost?" He then proceeded to conduct an elaborate cost-benefit analysis evaluating the replacement of a single standard urinal with a waterless-style urinal. He concluded that the replacement would save the university $6,000 a year. He received an A on his paper from a faculty member and then presented it to a packed room of facilities personnel. Following the presentation, we later determined that the eighteen toilets, urinals, and associated sinks and other water-using equipment in the building did not, in sum, use $6,000 of water annually! The student had not done any back-of-the-envelope estimation to ensure that his work was grounded in reality, had not validated his assumptions with facilities staff, and the faculty member overseeing the project was clueless. In an amusing e-mail exchange with TCI following this discovery, the student replied, "Good luck with this, I'm swamped with other papers and will leave for the summer in several weeks." Unfortunately, this type of interaction can leave busy facilities staff with the feeling that their time has been wasted, and can undermine their willingness to work with students again.

Students and Problem Analysis Often faculty have expectations about the analytic approaches that students should use in their projects. Faculty should ensure that students understand the requirements of the analytic procedure they are conducting, and its limitations. If, for example, students are expected to conduct a multiple regression analysis, both the students and the client should be aware that this is the case. The client may not be particularly interested in the results of the analysis, and may be much more interested in the raw data the students generate. Alternatively, the client may be very interested in the analysis and have concerns about the quality of the data. Although a client may be extremely familiar with analytic approaches required by the faculty or commonly used in the field, both faculty and students need to take the initiative to determine the level of client knowledge. Part of early planning discussions with the client may include gathering a sense of the extent to which the faculty can rely on the client to help students understand the limitations of the analytic approach they are using.

Although there is an occasional project in which the students realize midsemester that they need to use an analytic approach for which they are not trained, careful planning on the part of both client and faculty will minimize this type of problem. In one example at Tufts, students studied the HVAC systems of a building where there were frequent complaints of uncomfortable temperatures. The students erroneously assumed that the air-conditioning was turned too low and the heat too high in various conditions. They failed to understand the existing mechanical systems and the retrofits needed would have, in both their cases, increased energy use in order to improve comfort. For example, overheated rooms during the spring season were caused by energy-saving measures to keep the chillers off for as long as possible, not by the introduction of heat into the spaces in question. Students and the faculty who oversee them should take care that their assessments are within the limits of their expertise. When a faculty member's expertise is challenged by a particular project, the results of student study should be validated by an expert who is familiar with both the technology and its application before the student work is disseminated and presented as fact.

In another course, a graduate student team realized that their project would be a great deal more valuable to the client if they used GIS (geographic information systems) tools. As it happened, none of the team members had GIS training, and it was too far along in the semester to change the team's composition. One team member was willing to train intensively so that the final product could include some basic analyses using GIS, and the other members of the team shifted their responsibilities in recognition of the added time that would be consumed in unanticipated training. While the project ended up being a positive experience for both the client and the students, a less mature and accommodating group of students might have used the absence of GIS skills as the catalyst for a meltdown.

Developing Alternative Approaches Beyond the very significant student learning inherent in an applied project, an immensely valuable attribute of student projects may be that they cause the client to think about the problem in an entirely new manner. Although the fresh eyes of students may make contributions to any aspect of the project, our experience suggests that discussions of alternative approaches may stimulate creative thinking among university decision makers. On the other hand, alternative approaches may be most difficult for students to formulate.

What do we mean by alternative approaches? We mean a range of possible solutions to the problem or a range of possible actions to take. At the level of large projects, many students are familiar with this concept from their studies of the National Environmental Policy Act (NEPA), requiring environmental impact statements for federal decisions that significantly affect the environment. In addition to the "do nothing" alternative, and the recommended approach, these statements typically present alternatives that have different timelines, ranges of impact, or variations in scale. In a typical environmental impact assessment, alternatives are analyzed and a rationale is offered as to why the preferred approach is superior.

But alternative approaches may take on a different dimension depending on the type of question posed. In helping students formulate alternative approaches to a problem, there often is a fine line between providing students enough information so that they develop useful and creative approaches, and providing information that predisposes students to solutions that are favored by the project proponent. Although there is no universal antidote to this dilemma, one virtue of the Internet and of inexpensive telephone service is that for many problems, students can easily expand their list of alternative approaches by learning how the problem is solved by other institutions. Another challenge particular to alternative approaches is that students may have difficulty formulating alternatives for university climate change action that are "realistic." Of course, what is realistic may be in the eye of the beholder, and certainly varies across institutions.

Depending on their disciplinary training, students may not have had much experience with applied projects or in formulating alternative approaches, and may need coaching from faculty or from the project client. A two-part approach can be used. The faculty member associated with the project can explain the development of alternatives in a broad context (perhaps even in association with explanation building, formulation of hypotheses, and validity, depending on the level of the course) and the project client can articulate any alternatives that have already been placed on the table or that have been eliminated entirely.

In addition to conducting research on how the problem has been solved by other organizations, the development of alternatives can be an opportunity to help students acquire skills in brainstorming and other techniques for working together as a team. The degree of emphasis on these techniques will vary depending on the learning objectives of the course. If the course is designed to have students gather data and use specific analytic tools, too much emphasis on group process may be a distraction. On the other hand, in an upper-level professional program, it may be very useful to create opportunities for students to learn about and practice a range of techniques for group interactions. The development of alternative approaches can be a rewarding focus for this effort, both from the student and the client perspective.

Recommending a Course of Action Recommending a course of action can be a very valuable part of a student project for both the student and the client. For the student, it may be unfamiliar and thus uncomfortable to have to select a single approach from a wide array of attractive alternatives. This is especially true when data are imperfect, incomplete, or cannot be disaggregated. For students who are particularly uncomfortable with the recommendation phase, it may be useful to emphasize that there is rarely a right or wrong answer. Indeed, what may be most useful to the project client is following the logic that led to the selection of the preferred alternative. If the recommendations are framed artfully, they will be useful to the client even if the client places different value on the decision-making attributes and comes to a different conclusion about which course of action is preferred.

An astute client may learn, for example, that from the student perspective, aesthetics is more important than reliability or cost or ease of use for a piece of equipment or a building feature. The university decision maker can then test this finding in a variety of contexts. Are these design students or are they from engineering? Do others (faculty, administration, trustees, alumni) also place a very high value on aesthetics in this decision? What are the long-term implications for operation and maintenance costs of making a decision that favors aesthetics over reliability in this particular case? Is it worth further investigation with a design team or an equipment manufacturer to determine whether technology options exist or are emerging, or whether some other configurations can be developed such that the trade-off between aesthetics and reliability is not so stark?

In other words, recommendations from student projects may not result directly in the recommended action being taken, but may have the very positive result of encouraging decision makers on campus to view the problem in a new light. Both students and clients should regard an outcome of this type as a very successful project.

Evaluating, Monitoring, and Reporting on the Completed Project Reporting on the completed project can take many forms depending on the learning objectives of the course. In some courses, students are expected to prepare a written report and to give an oral presentation. If there is an identifiable client for the project, it can be very positive to invite the decision maker to be present in the audience for the students' oral presentation. Although students' anxiety level may skyrocket, the benefits can be significant in terms of providing experience in making presentations to a knowledgeable and interested individual, and in fielding questions about one's work. We have discovered that some busy clients are more likely to attend a student presentation than read a final paper. Nonetheless, a copy of the paper should be sent to the client with thanks.

When student projects involve development of action campaigns or informational materials for subsequent implementation or use by others, it may be important to students' sense of accomplishment if they understand what plans (if any) are in place to use their work. This will, of course, vary depending on the nature of the project. For example, a student project that examines other universities' development and implementation of cogeneration may inform the university's long-term strategic thinking. Other projects may inform shorter-term decisions, such as a decision to install VendingMisers on cold-drink machines. Even with tangible short-term outcomes, it has been our experience that students frequently graduate before their projects are implemented. TCI follows up with students on the implementation of their work when we can, and feel this could be an important part of the long-term relationship between an institution and its alumni.

Also, it is good practice to secure students' permission to post reports on the web, as TCI does for projects that we feel may be useful to decision makers in other institutions. This has the benefit of creating a "web publication" for the student, and facilitating the transfer of knowledge.

Monitoring the implementation and effectiveness of projects is a challenge at several levels. Good evaluation takes time and money, and when resources are scarce, it may be tempting to invest in new actions rather than evaluate the actions that have been put in place. On the positive side, some of the climate change action projects are very amenable to evaluation. The evaluation of Schmalz House in chapter 9 is an example of an informative assessment by a student. On the other hand, not all of the projects undertaken are comparably amenable to monitoring and evaluation. Few buildings at the university are submetered, so that it may be difficult to identify the effect of changes designed to reduce energy consumption. Even when there is individual metering, it may be difficult to discern the effect of some climate change actions, so calculated estimates or surrogate measures can be used. For example, a student experiment to test a social marketing campaign aimed at having residents of a large dormitory turn off computers when not in use did not reveal measurable changes in demand for electricity (see chapter 10 for details). This was not surprising in view of the small size of the target group. But the student was able to identify changes in knowledge and attitude, a finding that suggests that the experiment is worth repeating on a more ambitious scale.[2]

Challenges Taking Climate Change Action into the Classroom

When we use the university as a learning laboratory, traditional student, faculty, and staff roles are transformed, particularly when there is a third-party client. Students become more actively engaged in learning and faculty exercise a different kind of control over the learning experience. Staff become an integral part of the process and often become the primary teachers. Focusing on climate change action may be most successful if the instructor envisions the action element as a consulting or research project rather than as a conventional course experience. To a great degree, this challenge is similar to managing any complex project. One begins with an understanding of the strengths and weaknesses of the available personnel, in this case, the students. This seems obvious in a consulting environment, but it may represent a real departure for a faculty member with a room full of students.

Projects for climate action have significant benefits for student learning, but may have limited benefit to university staff. While the discussion above is presented to minimize the extent to which this occurs, in our experience great care needs to be taken to ensure that there is a net gain for staff. Students benefit from dealing with real problems, real data, and real limiting conditions. For some students, this will be their first experience in dealing professionally with other adults who have authority to implement changes they might recommend. This can be a highly motivating experience that provides transformational learning.

The next significant challenge is establishing a realistic set of tasks and a timeline for the project. This step is where several of our projects have run into difficulties. It may not be obvious how to break a complex project into the right number of pieces for a class. And further, it may be difficult to identify tasks that are equal in degree of difficulty, interest, glamour, and other attributes that are important to students. The easiest projects to carry out are those with greatest flexibility. The cost-benefit analysis project mentioned previously is an example of a very flexible project, because the number of analyses can easily be expanded to fit class size and the degree of complexity can be modified to reflect an individual or team effort.

Further challenges are introduced if student projects are undertaken in teams. Although we have taught many courses in which teamwork is a central learning objective, we still struggle to identify the ideal set of conditions for this vitally important and potentially difficult undertaking. The challenge is enhanced when TCI is the client for student work. Under what circumstances do we "stack the deck" and give the most important element of the project to the strongest team? When a climate action project is undertaken in one of our courses, we have someone else act as client so that we will not be tempted to influence the outcome of "our project."

Timelines for climate action can be an enormous challenge. The inescapable reality for classroom projects is the academic calendar. Projects must be specified in a way that allows for student completion and faculty evaluation on a schedule that meets the requirements of the institution. This is where the consulting analogy breaks down and projects are modified to comply with academic constraints. One approach to managing this situation is to require students to produce several interim products. Project outlines, multiple drafts, raw data sets, status reports, and other subtasks may be specified in an effort to ensure that student efforts are on target and on time. Otherwise we find that students may treat the project like a last-minute term paper and produce disappointing results.

Our experience suggests that many university decisions related to climate change action are not suitable for classroom projects. Timing is an important source of potential incompatibility. Need for specialized expertise and the complexity of energy systems and university business procedures are also factors. Most successful classroom projects are conceived the semester before they are undertaken. This means that projects are conceptualized and planned approximately six months before the information they generate is needed to inform decisions. This does not sound like a particularly daunting requirement, but it can be challenging in practice.

In working with university decision makers who are potential clients for student work, we are careful to manage expectations. No one can guarantee a priori that a student project will be high quality and will be focused appropriately to inform real-world decisions. Many projects meet or exceed expectations, but some do not. Most experienced decision makers have encountered situations in which paid consultants have produced disappointing products, so it is usually understood that a good outcome cannot be guaranteed. On the other hand, because the likelihood of student-project disappointment is perceived to be higher, it may make strategic sense to talk through an alternative approach to securing information at the time the decision maker agrees to become a student-project client.

Whether decision makers are amenable to using the university as a learning laboratory also may depend on the distribution of resources. We are "fortunate" in having an energy manager with inadequate staff resources. To the extent that projects can be conceptualized and developed to add value to the energy manager, we have a receptive audience for student work. On the other hand, we struggle with the need to protect the energy manager's time. The energy manager is not paid to educate students, nor can the university really afford to have her play this role. So TCI tries hard to craft projects that benefit the energy manager without requiring her to provide students with extensive briefings or to generate new data reports.

A final set of challenges associated with climate change action projects relates to time and other resources. We mentioned earlier the need to think carefully about whether the scope of a project is suitable for a semester project and attuned to student skills, but there are other challenges associated with time. Students may need to be reminded that their clients, especially if they are members of the administrative or operations staff, probably have something approximating a 9-to-5 job, whereas student time extends deep into late-night hours and often excludes the morning. This will mean scheduling client meetings during the "normal" workday and may prove challenging, particularly if students are working as a team. A practical solution (for which students may need emotional preparation) is to schedule early-morning client meetings. Often these time blocks are less likely to conflict with classes, study groups, and sports schedules. Time management by students is an important skill to teach and if done effectively, contributes to program success.

Demands for resources to conduct climate action projects will vary. Often students will want to travel, purchase information, make international telephone calls, collect and analyze samples, or produce a costly final report. It is important to be clear with students at the outset whether any of these expenses can be reimbursed, and if so, under what conditions. Some courses give students a very modest budget (approximately $100 per group) and reimburse expenses only when appropriate documentation is provided to the department sponsoring the course. In other cases, the client may establish a budget and rules for expenditures. Our experience is that it is not necessarily a problem for students in the same course if one client establishes a generous budget and another client does not, as long as the class as a whole is aware of these differences at the outset.

Climate action projects may place demands on campus resources that differ from those instructors are accustomed to using. For example, if there will be a class or classes at the end of the semester in which students make project presentations and clients and other guests are invited, it may be necessary to arrange for a room larger than the normal classroom. On our campus, competition for large rooms wired for projection capability is intense at the end of the semester, so anticipating the need for this type of resource will pay off. In addition, students may require resources not normally associated with the course. For example, a group may need access to the GIS lab and the plotter. If access is challenging for students not enrolled in a GIS class to negotiate, faculty should either anticipate the possibility and make arrangements in advance, or should advise students at the beginning of the semester that they see securing this type of access as a faculty responsibility and to ask for help when needed. Another group may decide that they wish to collect environmental samples and have them analyzed. Again, this is a case in which advance planning on the part of the client and faculty may facilitate access to laboratory supplies and analytic equipment. For projects only a semester long, the likelihood of a good outcome is increased if students know from the beginning what resource limitations they face and what resources can be made available, and plan accordingly.

Suggestions for Getting Started

Both faculty and students may assume, often incorrectly, that their climate action project idea is unique and has never been tried before on campus. A solution to this problem and a very constructive way to get started is to create an institutional memory for campus sustainability or climate change action projects. The exact form that such a memory might take will vary from institution to institution, and is limited only by the resources and creativity of the people involved in creating the memory.

An institutional memory is necessary, at least on our campus, because there is no formal mechanism across the institution for tracking and archiving student projects. After fifteen years of campus sustainability efforts, there have been several hundred student projects. Master's theses and doctoral dissertations are of course tracked and archived; however, many of the projects that inform climate change are generated as part of individual courses. Our Department of Urban and Environmental Policy and Planning has been the source of several climate change action projects through its core course called Field Projects, and the department routinely archives students' final reports; however, this is an exception. Many valuable student projects have been generated by senior engineers in their design course and by others all over campus. Whether final reports from these efforts have been retained may depend on the inclinations of individual faculty.

One way to get started on developing a memory is to create a student project in which key faculty and staff informants are interviewed and asked to provide information about projects in which they have been involved, as well as to provide materials that can reside in an archive. This project can be a useful activity in a social science course on qualitative research methods, and can give students practice in conducting interviews and organizing and coding interview results. Using the snowball approach, key informants can be asked to provide information on other projects that have been conducted. Students may well find that there are "urban legends"—perhaps more appropriately called "campus legends"—about projects that many people believe have been undertaken, but that cannot be verified through a rigorous process. On the other hand, students may also be astonished to learn the number of energy projects that have been conducted over the years, particularly starting in the 1970s, before most of them were born. Because the faculty sponsors of these early projects may be retired or on the verge of retirement, developing a memory sooner rather than later is a good strategy.

At a minimum, a memory itself can consist of a database of project titles, abstracts, and contact people that can be searched by keyword, date, faculty, and student. But a web-based multimedia approach offers the possibility of access to documents, videos of student presentations, links to related resources and to similar memories at other institutions, and a great deal more. If the memory is created as a strategy for getting started, a plan also should be developed for ensuring that it continues to meet needs as the climate action program matures.

Another way to get started, as we suggested earlier, is to decide on the most appropriate goal for the climate change action program. The goal may be replaced or amended in response to outside events or as program experience is gained, but we are strong advocates of starting with a numerical goal so that progress toward the goal can be evaluated periodically.

Students in Long-Term Projects and as Change Agents

Although we have talked extensively about projects that can be accomplished in a semester, some projects require more intensive effort over a longer period. In some cases a master's thesis or doctoral dissertation might satisfy both student and organizational goals, but here we can run into conflicts between student interests and faculty visions of what constitutes appropriate academic work in the discipline. Master's and doctoral work in interdisciplinary programs may expand the range of projects considered suitable to inform climate action.

Another approach is to plan climate action projects from the outset as having contributions from several courses and/or several generations of students. This is a particularly fruitful approach to ensuring that there is long-term monitoring and evaluation of projects, but rarely can it be left in the hands of individual faculty members for a variety of practical reasons. An organization such as TCI is ideally positioned to know which student projects have been selected for implementation by the university's operations staff, and to monitor and evaluate data over time through subsequent student projects. The projects for many courses that have supported Tufts climate change action focus on one set of skills—for example, designing an engineering project or conducting cost-benefit analyses. When the time is right, TCI can identify a set of course teaching skills associated with the next phase of the project, such as program evaluation or survey design and analysis. TCI staff can then approach the instructor of the program evaluation course and ask whether evaluating a particular climate action project on campus might be consistent with course requirements and student interests.

Links with Faculty Research

Research related to climate change can be conducted in any discipline, and is limited only by the imagination of the researcher and the availability of funds. Despite the failure of the United States to ratify the Kyoto Protocol, significant funds are available for a wide range of research through the U.S. Climate Change Science Program.[3] Many of the subelements of the program emphasize interdisciplinary research.

There is also funding available from some state agencies, private foundations, and corporations.

Articles related to climate change are appearing in a wide range of peer-reviewed journals, and journals focusing specifically on climate change are fast emerging. Perusing the reports of the IPCC provides an excellent overview of past and current peer-reviewed research and many ideas for further work. This is an area in which serious academic inquiry is emerging rapidly. Although traditional faculty research may address a nearly unlimited number of questions related to climate change, it is our sense that faculty work that informs climate action is still in its infancy.

Adjusting the scale of projects to accommodate the constellation of faculty interests, sponsoring-agency requirements, and campus climate actions may be a challenge. For example, a group of faculty at Tufts were successful in landing an early competitive grant to assess the impacts of climate change on the metropolitan Boston area.[4] The results of this fascinating project were presented to decision makers in several forums and were extremely well received. But the funded research stops short of being useful to inform the types of decisions faced by people who can take action to mitigate or adapt to the effects of climate change. For instance, decision makers are left not knowing what the implications might be for their campuses or communities, given significant differences in proximity to the shoreline, age of infrastructure, proximity to rivers and lakes, source of water supply, transportation features, source of energy supply, and many other considerations. A strategy to overcome this gap from knowledge to action is to have students focus senior projects or master's theses on follow-up questions, many of which may examine implications for campus action. We have had two graduate students use this project as a point of departure in examining effects on specific communities in the Boston area.

The faculty members who conduct the original research for a project may not have the time or inclination to move from their basic research to its application at a community or campus level. This is where good collaborative working relationships across campus are invaluable. Handing projects along to different groups of researchers and students as the core questions change and the research becomes more applied can be an exciting way of building a sense of community on a campus at the same time that individuals pursue projects they find most rewarding.

Summary

Climate change can have a place in many academic learning environments; however, faculty may need support to arrive at solutions suitable for their discipline. Institutions committed to climate change action may find it very rewarding to launch a faculty development program focused on climate change along the lines outlined in this chapter.

We focus most of our discussion on projects that support climate change action on campus because our experience suggests that they may pose challenges beyond those normally associated with developing and delivering a more conventional course. However, we feel their value to students, faculty, and the institution amply justifies the extra effort. Although we do not address in detail courses that serve to raise awareness and understanding of climate change (as distinct from informing climate change action), we believe their value is immeasurable. If all of our students understand climate change basics, we may help to create a generation of people who examine their own choices and who demand that governments and companies make choices in favor of emission reductions.

12

Degrees That Matter

Colleges and universities can and should take the lead in the full range of climate actions from improving the efficiency of operations to inventing technological solutions for governments, businesses, and individuals. Colleges and universities are communities. Like cities and towns, colleges and universities vary in size, resource use, demographics, and values. The challenge is to conduct activities in a way that minimizes harm. Growing food, manufacturing products, using and consuming goods and services; creating, heating, cooling, and providing energy services to buildings; and transporting people and goods all generate climate-altering emissions. This was true before the industrial revolution and it is true today. We face a climate change problem in part because efficient use of fossil fuels has not been a societal priority, nor has it been a priority to develop alternatives. This is changing. Outside the United States, a variety of ambitious national efforts are being taken to shift away from reliance on fuels with the greatest global warming potential, in some cases toward renewable sources such as wind. Within the United States, actions are being taken at the regional, state, municipal, and institutional level. Colleges and universities can inspire, inform, and enhance climate action at all levels.

The climate actions we encourage are not costly gestures designed simply to make a statement. Many of the measures we describe will increase efficiency, which means delivering the same energy service and performance with less energy. Other actions we discuss involve more systematic changes, such as switching to fuels with lower global warming potentials, increasing reliance on green power and alternative fuels, and reducing demand through changed behavior and expectations. Depending on local power markets, it may be possible to negotiate a contract with an alternative energy provider to deliver electricity at a rate that protects the institution from price increases more effectively than agreements with conventional power sources would. Distributed generation using an energy-efficient system may reduce carbon dioxide emissions and also may increase system reliability if the local power grid is vulnerable. Adaptation measures will protect assets and reduce the risk of damage from extreme weather events. Although we advocate for climate action because we have a social-responsibility agenda, we also support these actions because they make good business sense.

Emission reduction efforts at colleges and universities can inspire actions far beyond the campus. Alumni, staff, students, and faculty can take climate-sensitive practices into other spheres, practicing emission reduction in their communities and homes. Alumni can become active citizens, taking a responsibility for climate action into their workplaces regardless of their profession.

As we noted in earlier chapters, the college or university may be a significant local source of employment as well as intellectual and cultural life, and it also may be a significant source of heat-trapping gas emissions in its host community. With this prominent position comes both power and responsibility to act. We have said a great deal about the responsibility to take emission reduction action, but we have said relatively little about using the power of the position in the political arena.

When the Bush administration released its draft energy plan in spring 2001, we were astonished that a national plan could ignore climate change, ignore opportunities for dramatic increases in efficiency, and ignore developments in renewable and other innovative energy-supply technologies. So we drafted a letter for signature by the president of Tufts, and he in turn invited other college and university presidents to join him in expressing concern and offering help. To our great surprise, the letter had over forty signatures within a week. At least one reporter who covered the effort did attempt to elicit an administration comment to no avail. Tufts received a form letter from the White House. This action did not change national policy, but it did send a signal. We think it is time to send more signals.

Influencing policy related to climate action is a role that colleges and universities should consider. Whether operating collectively or as individuals in their fields of expertise, academics can make contributions to policy development. Many faculty members participate in the IPCC, serve on the many National Academy of Sciences committees that address climate change and related issues, and conduct research that informs policy development. If policies are modified to dramatically increase the fossil-fuel efficiency of the economy, then national emissions will be reduced, and colleges and universities will have a wider range of products from which to choose. At the state and local level, climate action plans are being developed and laws and regulations can be modified to greatly improve building performance. Through research and publications, academics can examine the many perverse subsidies embedded in our national energy policies that keep fossil-fuel prices artificially low and make climate-altering gas emissions unnecessarily high. Academics can propose policies and programs that align policy incentives with greenhouse gas reduction and address a host of related concerns, including improved fuel security and increased global equity. But people who operate regularly in the policy realm know that conducting sound scientific research and proposing thoughtful legislation are insufficient alone to ensure policy change. We encourage collaborative action to influence the policy process at the same time that we advocate taking action to reduce our own emissions.

Many of our homes, vehicles, campus buildings, and manufacturing facilities were designed and built when energy costs were relatively low. A variety of factors—including periodic energy price spikes, concerns about the environment and long-term oil supplies, and political crises such as the oil embargo of the 1970s—have all helped motivate the development of renewable energy technology and very efficient equipment. Yet, for a variety of reasons, these technologies have, until recently, largely failed to enjoy widespread mainstream application. Decades-old technologies for generating electricity and providing heat with fossil fuel remain in place, even though few of us would consider keeping computers or telephones with decades-old technology. Why do we treat energy differently? To some extent we must blame ourselves for this situation. If we walk into a car dealership and fail to ask which model is most fuel efficient, and fail to factor that information into our purchase decision, we become a data point supporting car companies' arguments that customers value performance over fuel efficiency. And so it goes for literally dozens of decisions that we make as individuals or as representatives of the organizations with which we associate. The sum total of these actions is that the United States is a profligate and inefficient user of fossil fuel and the world's largest emitter of climate-altering gases. Steps taken to reverse this trend will reduce emissions and, in many cases, save money.

For most institutions, governments, and many businesses, the greatest source of heat-trapping gas emissions is burning fossil fuel to heat and cool buildings, to generate electricity, and to transport people and goods. With widespread and strategic implementation of energy-efficiency measures, functionality of energy services can be maintained or improved and operating costs and emissions can be reduced.

Our university has proven to be an excellent environment for testing ideas about emission reductions because there are ample opportunities for increasing energy efficiency. If our operations were extremely energy efficient in 1990 and relied little on fossil fuels, then of course, achieving the reduction targets in the Kyoto Protocol would be very difficult or perhaps impossible (particularly if we were to add more square feet of built space). But our physical plant was not extremely efficient then, and like many other institutions, our infrastructure and our patterns of use still have room for improved efficiency. We have increased the amount of alternative energy in our mix, but we still rely primarily on fossil fuel. We have made some changes in the personal actions of the campus community, but more can be done. In this regard Tufts is a good model because it is typical. If it had been very easy to reduce greenhouse gases on our campus, we would not have had much of a story to tell.

Often we are asked to identify colleges and universities that are doing a good job on climate action. This is difficult. As we noted in chapter 2, comparative metrics, such as emissions per square foot or emissions per student, can be calculated; however, the list looks very different depending on which metric is selected. A campus can have several high-performance buildings, perhaps certified LEED platinum, and still fail the test of "doing a good job" if the buildings are far larger than needed, or if other buildings on campus are very inefficient. More conceptual work is needed to develop an equitable and meaningful ranking scheme that facilitates comparisons across institutions. The institutional research community is taking on this challenge. Until these metrics are developed, we believe it is constructive to think of a minimum credible climate action program that can apply to any college or university.

Attributes of a Credible Effort

In the previous chapters we talk about the amenability of different decision makers to change, and we acknowledge that climate actions feasible at one institution might be very difficult or infeasible at another. This is a recognition that having a champion for emission reduction is important, and it is also an acknowledgment that the culture of colleges and universities varies dramatically, especially in the extent to which sustainability is embraced by the institution. Our data indicate that there is a relationship between wealth and emissions of greenhouse gas, and a relationship between wealth and the ability to launch ambitious systematic approaches to energy efficiency and commitments to green power. All these factors aside, we believe there is a minimum set of attributes for all colleges and universities claiming serious climate action or claiming campus sustainability:

- A master plan that includes sustainability is in place. The plan itself or related planning activities address climate change mitigation and adaptation and develop an energy strategy that includes evaluation of alternative power and heating and cooling systems.
- All major buildings, especially those designed and built since 1990, are on energy management systems.
- Energy efficiency and emission reduction strategies are incorporated in university operations policies.
    - A funding mechanism for investing in on-campus energy efficiency.
    - Standards for energy efficiency in new buildings, renovations, and for equipment used on campus.
    - No incandescent lighting in classrooms, meetings rooms, hallways, or other public spaces. An active program is in place to replace the remaining incandescent bulbs, such as task lights in offices and residence halls.
    - No T-12 fluorescent tubes in ceiling fixtures. These are replaced with more efficient T-8 or T-5 lamps and reflectors, or the entire room lighting system is redesigned for greater energy efficiency.
- A robust recycling program is in place with convenient, functional, and well-marked containers in hallways, classrooms, dining areas, offices, and residences and at special events.
- The institution has a curriculum that creates opportunities for students to learn about climate change and climate action.
- A carpooling program is available for faculty and staff, as well as for students if there is a significant commuting student population.

• The institution has a process for identifying appropriate local goals such as water availability and soil erosion and evaluates progress toward those goals.
• Other criteria include a baseline inventory, an emission reduction goal, and periodic evaluation to measure progress in achieving the goal.

We hope that years from now, readers will look back on these minimal criteria and laugh because they are so modest. But our experience suggests that on some campuses, starting with a modest approach may be a pragmatic way to help decision makers gain confidence and to realize that emission reductions can be consistent with other community goals such as decreased operation and maintenance costs.

Here is a set of actions that will facilitate progress:

Set a Quantitative Goal and Measure Progress against the Goal

This is a familiar approach to capital campaigns, and many organizations use thermometer-like displays showing progress toward fundraising goals. Using this strategy for a climate change goal has all the same positive outcomes and perhaps a few more as well. If background work is done, it should be possible to understand the range of alternatives for meeting emission reduction goals and to have estimates of their costs. This may be an ideal opportunity to raise the awareness of the campus community about the relationship between decisions made daily and emissions of heat-trapping gases. Other approaches may emerge from community dialogue.

At the genesis of our program, we found the emission reduction goals of the Kyoto Protocol to be an attractive hook for students, faculty, staff, and sponsors. In spring 2003, our president raised the bar by pledging to meet the more ambitious goals established by the New England Governors and Eastern Canadian Premiers. We encourage research into goals of nearby institutions, cities, states, and regional groups. Working collaboratively with other organizations may have benefits beyond compatible goals. These benefits may include information sharing, technical assistance, positive presence in the media, and opportunities for grants.

What if we fail to reach the goal? That is always a risk, whether the enterprise relates to fundraising or the environment. With an emission reduction goal, people can learn a great deal in the process of "failing," and learning is, after all, the core mission of an academic institution. But the goal of climate action is to reduce net emissions of heat-trapping gas to the atmosphere. If we fail to do that, we fail in a critical stewardship mission.

Consider People and Technology as Systems

Some of the most challenging problems we have encountered are related to the interaction between people and technology. On our bad days (and there have been several), we have muttered through clenched teeth, "This is a people problem, not a technology problem." On our good days, we recognize that problems involving people and technology are more constructively thought of as problems with systems that can be fixed. Here is a very small example.

Walking around campus with the Tufts energy manager, we noticed that several recently installed compact fluorescent bulbs were missing from the lobby of Cabot Hall. This is the main building of our prestigious Fletcher School of Law and Diplomacy. Anyone could have taken the bulbs, but we assumed, probably unfairly, that students were the culprits. Could these students, the future diplomats and ambassadors of the world, many of whom are sent at great expense by their governments, have done something as undignified as steal lightbulbs? Possibly. The culprits could have been faculty or staff or visitors. Perhaps the thefts are an indicator of our success. We can take pride in the fact that members of the community were educated enough to recognize a compact fluorescent bulb when they saw one, and smart enough to know that if they used them in their apartments, they would save on electric bills. But how do we keep the next generation of students (or others using the building) from stealing the next generation of bulbs? The energy manager quickly pointed out that the bulbs were vulnerable because they were the screw-in type, and because the fixture is low it is easily accessible without a ladder. When we convert accessible fixtures so they are hardwired to accommodate plug-in compact fluorescent bulbs, our problem should disappear. The plug-in bulbs will only work in specially designed fixtures.

Value Incremental Progress

Our first new building project was the Wildlife Clinic at the School of Veterinary Medicine. We entered into that process "too late" for some of the energy-efficiency and emission reduction measures we wanted.

Some of the suggestions were deemed too expensive given budget limitations, even though the payback fell within the five-year period that had been agreed. We were disappointed because we knew that opportunities for efficiency lost at the time of construction are lost throughout the entire operational life of the building. As it was, our accepted recommendations saved energy, but the percentage of savings could have been higher if all our recommendations had been implemented. And greater savings would have been realized each year of the building's operation.

Did we succeed or did we fail on this project? Among TCI, the opinions varied. Students were inclined to the view that we failed because we fell short of what could have been, whereas faculty and staff took the view that we succeeded in introducing several major energy-efficiency measures that would not have been implemented without our intervention. We have been invited to participate in decision making related to subsequent new buildings, and through the Wildlife Clinic experience established a precedent for commissioning buildings. Here are some lessons related to new buildings:

- New-building construction is a high-stakes enterprise for an organization. There will be many constraints and many interests, some of which are in conflict.
- Participatory processes for vision, conceptual development, and design of new buildings are extremely valuable, but they are time consuming. At least in the environment in which we operate, we cannot expect to effect change if we voice an opinion once and walk away.
- A great deal of expertise is needed to carry out a vision for designing, constructing, testing, and commissioning a building that minimizes carbon dioxide emissions. Unless this expertise is available within the organization, it will be necessary to hire experts to participate in the process. Despite what they may claim, few architects, engineers, and project managers are experienced in the intensely collaborative and specialized work that needs to be undertaken to produce a very low emission building.
- Student engagement needs to be considered. Some colleges welcome student participation in new-building deliberations as a matter of course, but others make decisions on a case-by-case basis to advise on program, not on energy-using systems. Students are future building users, they perceive the nature of the community differently from faculty and staff, they have terrific ideas, they can work very hard (except when they have exams and papers due), and they have high expectations. Students are

inheriting a severely climate-altered world unless action is taken now, so they are critical stakeholders. On the other hand, engaging students can have drawbacks. If interested students lack experience or expertise, it can be very time consuming to provide the tutorials needed for effective participation. This tension increases when there are several projects on campus and when resources for staff or faculty mentoring are already stretched. One way of increasing the positive value of student engagement in new buildings is to incorporate the experience in a course or internship.

• The academic community needs to reconsider how we conduct research and teach students about the conceptualization, design, construction, and use of buildings. Opportunities exist for much greater attention to interdisciplinary work, collaborative processes, and reframing notions of successful outcomes. As the entire spectrum of building issues is scrutinized, perhaps the academy should consider reappropriating aspects of building design and operations that have been marginalized off campus and into the realm of practitioner knowledge.

In the new and existing building projects in which we have participated, there inevitably comes a point at which we convince ourselves that we have done the best we can. In other words, we embrace incremental progress. In part this is a coping mechanism. In part it is a recognition that we are members of a complex community with multiple goals for each building. In part it is an acknowledgment that we are not going to change the organization's decision making through a single project. We know that it may take several experiments with new approaches before they are modified and absorbed in ways that are congruent with our university's decision-making culture. It remains important to recognize that incremental innovation is a strategy that has proven successful in many organizations.

Appreciate the Challenges Associated with Change

TCI is acting as an internal agent for change, a champion for considering climate-altering gas emission reductions in all aspects of the university's decision making. But it seems unlikely that a transformation this broad can be achieved through the actions of a single change agent. This is why we engage the efforts of staff, students, faculty, and alumni.

There is no question that external pressure helps as well. These pressures can come from many sources. For example, the Kresge Foundation "has launched a national Green Building Initiative which provides planning and bonus grants, as well as educational information, for non-profit organizations interested in sustainable building and design."[1] Such an effort may motivate colleges and universities seeking Kresge support to also consider green buildings. In addition, government incentive programs, such as the Massachusetts Technology Collaborative's Renewable Energy Trust, are supporting renewable-energy features on buildings in the state.[2] Governments can also send supportive messages in other forms—for example, EPA's Energy Star program identifies efficient appliances, heating and cooling equipment, home electronics, lighting, office equipment, and commercial food service equipment, and establishes criteria for energy-efficient buildings.[3]

Several bodies of literature address aspects of change, and they may be useful in informing the planning, development, and implementation of climate action efforts. Our experience suggests that some of the barriers to change are internal to the organization and include:

- *An absence of rewards for experimentation and entrepreneurial activity.* This may seem astonishing for an academic institution and we would like to be proven wrong. However, it appears to us that fear of failure and a high degree of risk aversion have been factors in decisions taken in relation to new buildings, renovations, and operations that result in continued investments in outdated fossil-fuel technologies at institutions of higher learning nationwide.
- *Challenges associated with collaborative efforts.* Emission reduction is an inherently collaborative effort and the rewards are perceived as diffuse and risks high. Within an organization, people have to be willing to focus on achieving a goal of emission reduction rather than traditional measures of success. If performance were evaluated on the basis of collaborating to achieve progress toward organizational goals, this situation might begin to change.
- *Ineffective communication of priorities.* Complex organizations such as colleges and universities have many priorities, some of which may appear to be in conflict when an individual is facing a real decision. For example, the institution may value environmental performance and it may value cost savings. If an individual is ordering a replacement exhaust fan, she may face two choices, replacing the fan with the same model, or purchasing a more efficient model that costs 15 percent more. Unless the college has been explicit about favoring energy efficiency, it is quite likely that the decision will be made in favor of the less expensive replacement model. The decision maker may be unaware that the more expensive and efficient model will cost less to operate, and in the long run, the

lower operational costs will make the efficient fan less costly for the college. To achieve energy efficiency, you often have to spend money to save. Particularly among parsimonious Yankees, this can be a challenging concept.

Barriers to change external to colleges and universities include:

• *Prices for goods and services that do not reflect their full environmental costs.* Although academics, including our colleague Neva Goodwin, have advocated for natural resource accounting and modified metrics to replace the GDP, limited progress has been realized.

• *Existence of policies that encourage fossil-fuel use.* Perverse incentives in U.S. law artificially lower the cost of fossil fuels, forcing renewable-energy providers to compete on a playing field that is far from level. State regulations may discourage distributed generation, and local zoning may discourage wind installations.

• *Absence of policies to encourage emission reduction.* Opportunities should be seized at the federal level to dramatically increase efficiency of vehicles and equipment. Financial instruments can be designed to encourage investments in efficiency. Large-scale research and development efforts in renewable energy can be funded.

Appreciate the Limits of Your Expertise

It requires some level of technical expertise to know how to identify and prioritize emission reduction opportunities. This book has illuminated many aspects of our learning process. And there is a vast range of resources that can help inform decisions. Once priorities have been established, it often requires specialized knowledge to develop plans and specifications for specific action that will reduce emissions. Many resources are available:

• The Internet is a valuable resource for getting started and for learning about the activities of other colleges and universities.

• There are specialized resources for new buildings, renovations, and many major decisions related to campus infrastructure.

• There are a wide range of experts in high-performance building planning and design, energy engineering, modeling, testing, and building commissioning who can be hired as consultants.

• Asking the right questions is essential to learning and to helping make good decisions.

Effective climate action advocates understand their limits and seek outside expertise frequently.

Embrace Systems Thinking

Colleges and universities are complex institutions that have many complex systems operating within them. Some of the systems that are of greatest importance in reducing carbon dioxide emissions include energy generation, distribution, and consumption; transportation; and buildings. Other heat-trapping gases are associated with coolants, landscaping and agriculture, waste disposal, and use of specialty chemicals. Some of the decision-making systems that influence these important physical systems are purchasing, financial management (especially allocation of overhead costs), deferred maintenance, master planning, sustainability policies and construction standards, and performance evaluations.

It is impossible for a climate change advocate to influence all of these systems at once, yet it is important to recognize that they are all related when the goal is reducing climate-altering gas emissions.

Consume Thoughtfully

Consumption and climate-altering gases are linked. There is no question that many necessary functions of the college can be made more efficient and can result in fewer emissions. At the same time, some institutions are riding a wave of construction and providing amenities that raise questions about needs as distinct from wants. Chapter 9 (box 9.2) provides examples of colleges and universities attempting to attract students with movie theaters, indoor kayaking streams, and massage parlors. While not all of these have obvious climate change implications, we take the view that climate change should be considered in all aspects of university decision making, but particularly with respect to the built environment.

Colleges and universities have enough local and collective purchasing power to help create demand for products and services that produce fewer emissions than the conventional alternatives. Whether this comes in the form of decisions to procure locally grown food, to sign a long-term contract for the campus to be powered by wind, or to purchase electric and hybrid vehicles for the university's fleet, the cumulative impact of these decisions will make a difference.

Educate the Marketplace

In the process of taking climate action on campus, we have had many interactions with the marketplace for climate action services. Some of these have been very satisfactory, but many have not. This leads us to conclude that there is room for educating those providing services so that taking climate action is more like taking a shuttle from Boston to New York than taking a trip to the moon. Below are some examples.

An early solar problem emerged when TCI was engaged with its initial demonstration project, renovations to Schmalz House. Specifications for the solar thermal system were developed with the assistance of a mechanical engineer experienced with solar systems. University procedures call for obtaining bids from three prospective vendors for purchases greater than $10,000, so the $14,000 solar thermal system needed to go through this process. Although it was easy to identify firms in the Boston area that advertised expertise in solar installations, it was surprisingly difficult to find firms willing to submit bids.

A graduate student working for TCI found that, during the summer of 1999, many solar installers did not return telephone calls, did not have answering machines, did not have active websites, and did not have working fax numbers. Finding bidders was a time-consuming and frustrating experience even for a person who regarded this as her priority assignment. A further challenge was finding solar installers that met the university's requirements for bonding and insurance. Installing heavy solar panels on a roof involves an obvious element of risk, and any property owner, not just a university, would be well served by requiring that the installer insure employees against accidents.

We talked internally about the apparent lack of business savvy demonstrated by this interaction with solar installers in the Boston area. While we could speculate at length about the reasons for the phenomenon, the implications were sobering. If solar installers are not sufficiently businesslike to meet basic requirements for interacting with the university, it comes as no surprise that solar use is still the exception. The house on which solar thermal panels were installed is in a neighborhood of nearly identical houses in a residential community at one edge of campus. So in many respects our experience was in effect a demonstration of what it is like for a homeowner to undertake a solar project. And it is no wonder that few homeowners take on the challenge!

As time has gone on, we have learned that our initial experience was not just bad luck, and instead is indicative of an important and frequently overlooked set of barriers to the diffusion and widespread use of solar technology. Although we sense that there are improvements all around in development and implementation of renewable-energy technology, we still see gaps. Colleges and universities can play a role in nurturing myriad aspects of this fledgling industry.

Another set of barriers to emission reduction in new buildings has been described as the problem of creating demand. To distill a complex set of interactions into a simple chain of events: architects design what the owner wants; engineers work with constraints established by the design and the technology in existing similar buildings; builders meet specifications established by engineers and architects and minimize costs. But owners generally do not walk into an architect's office saying, "I want a building that will generate no net emissions of climate-altering gases." This situation is changing. Relentless efforts on the part of dedicated individuals and organizations have created attractive buildings that use a minimum of fossil fuel with competitive construction costs and with operating and maintenance costs that are substantially less than conventional costs. Such efforts require, among other things, close collaborative efforts among the entire design and construction team. Colleges and universities must take the lead in creating the demand for this type of building.

Build Alliances for Action

Particularly in a college or university where faculty and students may have interests or expertise in an area related to climate change, it may be easy to overlook the implications of the simple observation that staff in the operations division take most of the actions when it comes to achieving emission reductions on campus. When faculty complain, "Why can't we just tell them to do it differently?", a useful response is, "How would you like to have someone from Facilities tell you how to teach microeconomics (or mechanical engineering or thermodynamics)?" That having been said, there certainly may be a need for workshops and other educational opportunities for decision makers ranging from trustees to operations personnel. TCI has found that paying for operations staff to attend green-building workshops and our planning and hosting high-performance-building seminars has had very positive results.

We advocate crafting projects so students can solve real problems with real data and can augment the resources of operations staff. We also recognize that without safeguards, project quality may be compromised by inadequate data, inadequate student understanding of systems, protocol, technology or finance, poor timing, or excessive demands on the time of operations staff to oversee the project.

The plain truth is that students and faculty may have brilliant ideas but may never have been in the position of moving from theory to action. By using the university community as a learning laboratory or clinic, more practice can be built into the educational experience and more resources can be focused on climate action.

What Really Matters: Acting Now

Climate action can and must be taken at many levels. Governments, companies, nongovernmental organizations, primary and secondary schools, and individuals all have important roles to play. But colleges and universities occupy a unique place in society, and their example and influence will have a disproportionately large and positive influence on climate action. Institutions of higher learning are sources of inspiration, creativity, and leadership within their communities. The increasingly creative and planful decisions of college and university administrators and staff, particularly in campus operations, can serve as a constructive model for infrastructure managers in other sectors. In educating students, there is an obligation to teach about an issue that will have such a significant influence during their lifetimes, and a great value in engaging their active participation in developing solutions. Students are vocal and passionate about causes in which they believe, and alumni are positioned to influence a widening circle of people, places, and organizations. Faculty conduct research at the center of climate change science, policy, and technology, but creating knowledge is insufficient. It is time for a concerted effort to transfer knowledge from the laboratory and the classroom into practice. Acting now is essential. Colleges and universities whose graduates are eager to create an equitable society that takes effective climate action will be granting degrees that matter.

Appendix A
Global Warming Potential

Global Warming Potential (GWP) is intended as a quantified measure of the globally averaged relative measure of the warming potential of a particular greenhouse gas. It is defined relative to a reference gas. Carbon dioxide (CO_2) was chosen as this reference gas. This means that carbon dioxide has a GWP of 1.

Global warming potential

Gas	Atmospheric lifetime[a]	100-year GWP	Uses
Carbon dioxide (CO_2)	50–200	1	Electricity, heat, transport, etc.
Methane (CH_4)[b]	12 ± 3	21	Veterinary school, compost, waste in landfills
Nitrous oxide (N_2O)	120	310	Anesthetic gas used in the dental school
HFC-23	264	11,700	Semiconductor manufacturing and as a fire-extinguishing agent
HFC-125	32.6	2,800	Stationary refrigeration and air-conditioning applications
HFC-134a	14.6	1,300	Automotive air conditioners
HFC-143a	48.3	3,800	Substitute for CFCs
HFC-152a	1.5	140	Blowing agent, an ingredient in refrigerant blends
HFC-227ea	36.5	2,900	Substitute for CFCs
HFC-236fa	209	6,300	Substitute for CFCs
HFC-4310mee	17.1	1,300	Substitute for CFCs

(continued)

Gas	Atmospheric lifetime[a]	100-year GWP	Uses
CF4	50,000	6,500	Byproducts of aluminum smelting and semiconductor manufacturing
C2F6	10,000	9,200	Byproducts of aluminum smelting and semiconductor manufacturing
C4F10	2,600	7,000	Byproducts of aluminum smelting and semiconductor manufacturing
C6F14	3,200	5,400	Byproducts of aluminum smelting and semiconductor manufacturing
SF6	3,200	23,900	Insulator for circuit breakers, switch gear, and other electrical equipment; a useful atmospheric tracer gas for a variety of experimental purposes

Notes

a. The lifetime of a greenhouse gas refers to the approximate amount of time it would take for the anthropogenic increment to an atmospheric pollutant concentration to return to its natural level (assuming emissions cease) as a result of either being converted to another chemical compound or being taken out of the atmosphere via a sink. This time depends on the pollutant's sources and sinks as well as its reactivity. The lifetime of a pollutant is often considered in conjunction with the mixing of pollutants in the atmosphere; a long lifetime will allow the pollutant to mix throughout the atmosphere.

b. The methane GWP includes the direct effects and those indirect effects due to the production of tropospheric ozone and stratospheric water vapor. The indirect effect due to the production of CO_2 is not included.

Source: "Greenhouse Gases and Global Warming Potential Values: Excerpt from the Inventory of U.S. Greenhouse Emissions and Sinks: 1990–2000," http://yosemite.epa.gov/oar/globalwarming.nsf/UniqueKeyLookup/SHSU5BUM9T/$File/ghg_gwp.pdf.

Appendix B
Information Related to Climate Change and Climate Change Action

There is a vast body of web-based information related to climate change and climate change action in university facilities. The appendix is designed to help you get started.

Climate Change Science and Research

Carbon Dioxide Information Analysis Center
http://cdiac.esd.ornl.gov/home.html

Center for Health and the Global Environment, Harvard Medical School
http://www.med.harvard.edu/chgc

Environmental Protection Agency
http://www.epa.gov/globalwarming
http://www.epa.gov/globalwarming/climatelink

Global Warming Basics
http://yosemite.epa.gov/oar/globalwarming.nsf/content/index.html
http://www.nrdc.org/globalWarming/f101.asp#1

Goddard Institute for Space Studies
http://www.giss.nasa.gov

Hadley Centre (United Kingdom)
http://www.met-office.gov.uk/research/hadleycentre/

Health Ecological & Economic Dimensions of Major Disturbances Program
http://www.heedmd.org

International Panel on Climate Change
http://www.ipcc.ch/

Kyoto Protocol
http://www.unfccc.de/resource/docs/convkp/kpeng.html

National Climatic Data Center (of the National Oceanic and Atmospheric Administration)
http://www.ncdc.noaa.gov

National Oceanic and Atmosphere Administration
http://www.noaa.gov

Pacific Institute
http://www.pacinst.org/wildlife.html

RealClimate (climate scientist–led blog)
http://www.realclimate.org/

U.S. Carbon Inventory
http://www.eia.doe.gov/oiaf/1605/flash/flash.html

U.S. Global Change Research Program
http://www.gcrio.org

Understanding Climate Change: A Beginner's Guide to the United Nations Framework Convention
http://www.unfccc.de/resource/beginner.html

UNEP/GRID-Arendal
http://www.grida.no/

United Nations Environment Program (worldwide environmental data)
http://www.unep.org/geo2000/english/index.htm
http://www.unep.org/geo2000/english/figures.htm

United Nations Framework Convention on Climate Change
http://www.unfccc.de
http://unfccc.int/essential_background/convention/items/2627.php

Woods Hole Oceanographic Institution
http://www.whoi.edu/

World Meteorological Organization
http://www.wmo.ch

WRI Issue Brief: Climate Science 2005—Major New Discoveries
http://pdf.wri.org/climatescience_2005.pdf

Climate Change Advocacy and Policy

Centre for Science and Environment (New Delhi, India)
http://www.cseindia.org

Climate Institute
http://www.climate.org

Climate Solutions
http://www.climatesolutions.org

CNE (Climate Network of Europe)
http://www.climatenetwork.org

Environmental Defense
http://www.edf.org

Fight Global Warming.org

Germanwatch
http://www.germanwatch.org

The Heat Is On
http://www.heatisonline.org

National Environmental Trust
http://www.environet.org

Natural Resources Defense Fund
http://www.nrdc.org

Pew Center on Global Climate Change
http://www.pewclimate.org/

Public Interest Research Group (PIRG)
http://www.pirg.org

Sierra Club Global Warming Campaign
http://www.sierraclub.org/globalwarming/

U.S. Business Council for Sustainable Energy
http://www.bcse.org

U.S. Climate Action Network
http://www.climatenetwork.org/uscan.htm

Union of Concerned Scientists
http://www.ucsusa.org

World Resources Institute
http://www.wri.org

World Wildlife Fund
http://www.worldwildlife.org/climate

Worldwatch Institute
http://www.worldwatch.org

Climate Change Planning

EPA Climate Change Action Plans
http://yosemite.epa.gov/oar/globalwarming.nsf/content/ActionsState
ActionPlans.html

ICLEI—Local Governments for Sustainability
http://www.iclei.org/
http://www.iclei.org/co2

Campus-Based Organizations

APPA: The Association of Higher Education Facilities Officers
http://www.appa.org

Association for the Advancement of Sustainability in Higher Education
http://www.aashe.org

Clean Air Cool Planet
http://www.cleanair-coolplanet.org

National Association of College and University Business Officers
http://www.nacubo.org

NWF's Campus Ecology Program
http://www.nwf.org/campusecology/

Society of College and University Planners
http://www.scup.org

Tufts Climate Initiative
http://www.tufts.edu/tci

Renewable-Energy and Energy-Efficiency Information

Alliance to Save Energy
http://www.ase.org

American Council for an Energy Efficient Economy
http://www.aceee.org

American Society of Heating, Refrigerating and Air-Conditioning Engineers (ASHRAE)
http://www.ashrae.org

Best Practices—DOE Industrial Technologies Program
http://www1.eere.energy.gov/industry/bestpractices/

Building Energy Software Tools
http://www.eere.energy.gov/buildings/tools_directory/

Center for Renewable Energy and Sustainable Technology
http://www.crest.org

Energy Efficiency and Renewable Energy Network (EERE)
http://www.eere.energy.gov

Energy-Efficient Lighting
http://www.eere.energy.gov/EE/buildings_lighting.html

Energy Star
http://www.energystar.gov

Environmental Building News
http://www.buildinggreen.com

FEMP Operations and Maintenance
http://www.eere.energy.gov/femp/operations_maintenance/om_best_practices_guidebook.cfm

Industry Plant Managers and Engineers
http://www.eere.energy.gov/consumer/industry/

International Association of Energy Efficient Lighting
http://www.iaeel.org

Laboratories
http://www.labs21century.gov/

MTC Renewable Energy Trust Green Buildings Initiative
http://www.masstech.org

Northeast Sustainable Energy Association (NESEA)
http://www.nesea.org/

Purchasing Specifications for Energy-Efficient Products
http://www.eere.energy.gov/femp/procurement/

Rocky Mountain Institute
http://www.rmi.org

U.S. DOE Energy Efficiency and Renewable Energy
http://www.eere.energy.gov/

Information on Green Buildings

American School and University magazine
http://www.asumag.com/

BuildingGreen.com (producers of GreenSpec, a specification guide and *Environmental Building News*)
http://www.buildinggreen.com

Combined Heat and Power Resources
http://www.eere.energy.gov/de/chp/chp_applications/information_resources.html

Computers and Information Technology Equipment
http://www.apcmedia.com/salestools/SADE-5TNRKG_R0_EN.pdf

DOE Distributed Energy Program
http://www.eere.energy.gov/de

Energy User News
http://www.energyandpowermanagement.com/

FEMP CHP Resources
http://www.eere.energy.gov/femp/technologies/derchp.cfm

Federal Energy Management Program
http://www.eere.energy.gov/femp/

Greener Buildings (from GreenBiz.com)
http://www.greenerbuildings.com/

New Building Institute, Inc.
http://www.newbuildings.org

Northeast Sustainable Energy Association
http://www.nesea.org

Oak Ridge National Laboratory FEMP
http://www.ornl.gov/sci/femp/index.shtml

U.S. Green Buildings Council
http://www.usgbc.com

United States Combined Heat and Power Association
http://uschpa.admgt.com/

Whole Building Design Guide
http://www.wbdg.org

Home Information

Buying Clean Electricity
http://www.eere.energy.gov/consumer/your_home/electricity/index.cfm/mytopic=10400

EERE Home Information
http://www.eere.energy.gov/consumer/your_home/

Energy Savers Booklet site
http://www.eere.energy.gov/consumer/tips/

Home Space Heating and Cooling
http://www.eere.energy.gov/consumer/your_home/space_heating_cooling/index.cfm/mytopic=12300

Information Resources
http://www.eere.energy.gov/consumer/information_resources/

Making Clean Electricity
http://www.eere.energy.gov/consumer/your_home/electricity/index.cfm/mytopic=10510

Your Vehicle
http://www.eere.energy.gov/consumer/your_vehicle/

Search Engines about Climate Change

Climate Ark
http://www.climateark.org

Keywords to help in searching for climate change information:

Climate change
Global warming
Greenhouse gas emissions
Energy efficiency
Energy policy
Emission factors

Appendix C
Elements of an Emissions Inventory

A campus or university inventory of climate-altering gas emissions is the sum of all the activities that produce greenhouse gas emissions multiplied by the global warming effect of each of these activities. The inventory is usually bounded by explicit limits. There typically include a physical boundary (e.g., a project, a building, a campus, a university), the time frame (e.g., calendar year, fiscal year, or months), and the specific activities that you elect to include:

The sum of (Activities × Emission factors) = Emissions

The same principles apply to a single energy-saving project that apply to the campus as a whole.

Activities

The *activities* are measured in a specific units related to the activity. Some examples include:

- Therms or cubic meters of natural gas burned
- Gallons of heating oil burned
- Kilowatt-hours of electricity purchased or generated
- Gallons of gasoline used
- Business miles or kilometers traveled

Activity data are most commonly measured as the quantity of a commodity that is purchased. In some cases it may be possible to measure the actual usage. For example, fuels are measured based on purchase date unless they are also metered as they enter a central boiler plant. Electricity purchases are equivalent to use.

Primary Activities

Primary activities have emissions resulting from fuels burned or other activities undertaken on campus. These include fuels burned for heating and cooling, fuels used for any on-site electricity generation, gasoline purchased, or fuels used in university vehicles.

Secondary Activities

In this case, emissions result from the combustion of fuels by a third party. These emissions are generated by the third party on the institution's behalf. Not all inventories include secondary emissions, and some double counting can result if both the generator and user count these activities. If the secondary activity would not occur if the institution didn't need it, it makes sense to count the activity. (For example, if the university reduces electricity use from energy efficiency efforts, it makes sense for the university to see a reduction in its emissions from secondary activity emissions.) Examples of secondary activities include:

- Electricity generated off site (typically included in a campus or university inventory)
- Steam generated off site and purchased (typically included)
- Transportation
  - Deliveries
  - Business travel
  - Commuting
  - Flying
- Goods purchased
  - Food
  - Building materials

The *Tufts Inventory* includes annual, universitywide activity data from

- #6 fuel oil (purchased)
- #2 fuel oil
- Natural gas
- Propane
- Purchased electricity
- Purchased steam
- Commuting

- Purchased gasoline and diesel
- Dairy herd
- Solid waste and recycling

The *Tufts Inventory* does not include activity data from

- Nitrous oxide
- Refrigerant leaks
- Air travel
- Secondary purchasing emissions
- Construction-related emissions
- Product-related emissions (e.g., embodied energy)

Our rationale for including particular data and excluding other data is based on the magnitude of the impact, the time required to gather the data, the availability of data, and the utility of the information.

Emission Factors

An emission factor converts *activity data* to emission values. Appendix D has the most common emission factors for converting fuels to pounds of carbon dioxide per million Btu or per unit volume or mass.

Determining emission factors for secondary sources (electricity and steam) is challenging and must include the input fuels and the efficiency of the system that provides the product.

Conversion Factors

Emission factors may be given in terms of pounds, short tons, or metric tonnes of carbon or carbon dioxide. Care must be taken to use the same units throughout the inventory and to use the same units in comparing results.

To convert short tons (2,000 pounds) to metric tonnes, multiply by 0.9071848.
To convert carbon dioxide to carbon, multiply by 0.2727 (12/44).
To convert other gases to a carbon equivalent, multiply by their global warming potential (see appendix A).

Note: The carbon-to-carbon dioxide conversion is based on the molecular weight of each element: C = 12 and CO_2 = 44.

Challenges

The concept of conducting an inventory is simple, but undertaking a universitywide inventory is, in fact, a complex task. We recommend starting with the very basic data from fuels burned and electricity purchased and adding elements as you have time and resources. Our experience suggests that there are numerous challenges. These include:

- The number of data points to collect
- The availability of data, especially for past years
- The method of data recording (most data is kept for billing purposes and may include only dollars and not quantities)
- Data estimation (some utility data are estimated, generally high, and adjusted on the next month's bill; the billing records may be adjusted, but not necessarily the usage records)
- Lack of metering to see the effect of specific projects

Appendix D
Emission Factors for Fuels

Fuel	Code	Emission coefficients	
		Pounds CO_2 per unit volume or mass	Pounds CO_2 per million Btu
Petroleum products			
Aviation gasoline	AV	18.355 per gallon 770.916 per barrel	152.717
Distillate fuel (no. 1, no. 2, no. 4 fuel oil and diesel)	DF	22.384 per gallon 940.109 per barrel	161.386
Jet fuel	JF	21.095 per gallon 885.98 per barrel	156.258
Kerosene	KS	21.537 per gallon 904.565 per barrel	159.535
Liquified petroleum gases (LPG)	LG	12.805 per gallon 537.804 per barrel	139.039
Motor gasoline	MG	19.564 per gallon 822.944 per barrel	156.425
Petroleum coke	PC	32.397 per gallon 1,356.461 per barrel 6,768.667 per short ton	225.130
Residual fuel (no. 5 and no. 6 fuel oil)	RF	26.033 per gallon 1,093.384 per barrel	173.906
Natural gas and other gaseous fuels			
Methane	ME	116.376 per 1,000 ft3	115.258
Landfill gas	LF	[1] per 1,000 ft3	115.258
Flare gas	FG	133.759 per 1,000 ft3	120.721
Natural gas (pipeline)	NG	120.593 per 1,000 ft3	117.080
Propane	PR	12.669 per gallon 532.085 per barrel	139.178
Electricity	EL	Varies depending on fuel used to generate electricity	

(continued)

Fuel	Code	Emission coefficients: Pounds CO_2 per unit volume or mass	Emission coefficients: Pounds CO_2 per million Btu
Electricity generated from landfill gas	LE	Varies depending on heat rate of the power generating facility	
Coal	CL		
Anthracite	AC	3,852.16 per short ton	227.400
Bituminous	BC	4,931.30 per short ton	205.300
Subbituminous	SB	3,715.90 per short ton	212.700
Lignite	LC	2,791.60 per short ton	215.400
Renewable Sources			
Biomass	BM	Varies depending on the composition of the biomass	
Geothermal energy	GE	0	0
Wind	WN	0	0
Photovoltaic and solar thermal	PV	0	0
Hydropower	HY	0	0
Tires/tire-derived fuel	TF	6,160 short tons	189.538
Wood and wood waste[2]	WW	3,814 per short ton	221.943
Municipal solid waste[2]	MS	1,999 per short ton	199.854
Nuclear	NU	0	0
Other	ZZ	0	0

Notes

1. For a landfill gas coefficient per thousand standard cubic feet, multiply the methane factor by the share of the landfill gas that is methane.
2. These biofuels contain "biogenic" carbon. Under international greenhouse gas accounting methods developed by the Intergovernmental Panel on Climate Change, biogenic carbon is part of the natural carbon balance and it will not add to atmospheric concentrations of carbon dioxide.[3] Reporters may wish to use an emission factor of zero for wood, wood waste, and other biomass fuels in which the carbon is entirely biogenic. Municipal solid waste, however, normally contains inorganic materials, principally plastics that contain carbon that is not biogenic. The proportion of plastics in municipal solid waste varies considerably depending on climate, season, socioeconomic factors, and waste management practices. As a result, EIA does not estimate a nonbiogenic carbon dioxide emission factor for municipal solid waste. The U.S. Environmental Protection Agency estimates that, in 1997, municipal solid waste in the United States contained 15.93 percent plastics and the carbon dioxide emission factor for these materials was 5,771 lbs per ton.[4] Using this information, a proxy for a national

(continued)

average nonbiogenic emission factor of 919 lbs carbon dioxide per short ton of municipal solid waste can be derived. This represents 91.9 lbs carbon dioxide per million Btu, assuming the average energy content of municipal solid waste is 5,000 Btu/lb.

3. Intergovernmental Panel on Climate Change, *Greenhouse Gas Inventory Reference Manual: Revised 1996 IPCC Guidelines for National Greenhouse Gas Inventories*, vol. 3 (Paris: Intergovernmental Panel on Climate Change, 1997), 6.28.

4. U.S. Environmental Protection Agency, *Inventory of U.S. Greenhouse Gas Emissions and Sinks: 1990–1998*, EPA 236-R-00-001 (Washington, DC: U.S. Environmental Protection Agency, April 2000).

Source: Energy Information Administration, http://www.eia.doe.gov/oiaf/1605/factors.html.

Appendix E
Sample Projects for Tufts Students

The Tufts Climate Initiative works within Tufts as an advocate and resource for helping Tufts to reduce greenhouse gas emissions from Tufts' own activities. TCI is interested in working with students on projects that are related to reducing the environmental impact on campus. The topics listed below are sample student projects. Most are designed for graduate students. Some assume that the student or student team will have some engineering expertise.

Feasibility of Methane Digesters for Use at Tufts

Tufts School of Veterinary Medicine's animals (cows and pigs) generate large quantities of manure. Cow manure is concentrated in dairy barns, and swine manure is collected in large tanks in a highly dilute form. This project will examine the feasibility of methane digesters for generating electricity. The project will answer questions such as

- What is the technology? Where else is it being used?
- How much manure is needed and how much electricity can be produced?
- What are the design considerations?
- Can manure management at Tufts be adapted to create a fuel source? What is needed to do so?
- What are the costs/benefits?
- How might funding be obtained for undertaking a project of this sort?

Wind-Power Feasibility

Wind-power has the potential to generate significant electricity nationwide. This project will examine the feasibility of new building-integrated wind-power technologies for use at one of Tufts' three campuses (Medford, Boston, Grafton). Questions to ask include:

- Is the technology feasible?
- Is there sufficient wind on any of our campuses?
- Where are there installations?
- What are the costs, design considerations, permits, and/or visual issues?
- How much power can be produced given the conditions here?

Investigation of Energy-Efficient Commercial Kitchen Equipment

Although energy costs in dining (estimated to be about 13 cents per meal) are only a fraction of dining's labor and food costs, the dining-services energy can amount to 10 percent of total university energy costs. Highly energy-efficient, Energy Star–rated appliances are readily available to consumers in the retail market. This project will

- Determine manufacturers and availability of energy-efficient commercial kitchen equipment
- Determine possible opportunities at dining facilities on the Medford main campus
- Identify case studies of facilities that have installed this technology and research how the technology is actually working
- Determine cost and benefits
- Identify installation and maintenance issues or benefits

Computers and Power Management

The number of computers owned by Tufts University and Tufts students has grown to about 7,300. Much of the increase in Tufts electricity consumption has been attributed to the increase in technology use. A study done by Tufts undergraduates in 1999 revealed that nearly 80 percent of students leave their computers sometimes or almost always on.

This project will conduct a new, more in-depth survey that will explore the current usage patterns of Tufts staff and students. The results of the survey will then be used to develop an action plan to minimize electricity consumption from computers.

It will also include research on the following topics:

• Flat screens: electricity consumption, LCA of monitors (e.g., does it take significantly more energy to produce a flat screen compared to a regular monitor)?

• Research and test with measuring device (line logger) actual energy consumption of several different computers (active, screen off, sleep mode)

• Research and update information on power management

Air-Travel Research

Most people do not realize the severe impact air travel has on climate change and air pollution. This project will research the impact of air travel and develop a brochure and web page with information, comparisons, and suggested alternatives for students, staff, and faculty.

Research will include the following topics:

- Climate change impact of high-altitude emissions
- Emissions of different types of aircrafts
- Impact of landing and take-offs
- Impacts of short-distance flights versus long distance flights
- Alternatives to air travel and offset possibilities
- Issues related to offsetting emissions by forest sequestration
- Number of annual flights in the United States/in the world

Solar Hot Water at the Hospital for Large Animals—Grafton Campus

The Hospital for Large Animals facility on the Grafton campus operates year-round and provides consultation, referral, and emergency veterinary services, as well as twenty-four-hour care for horses, cattle, sheep, goats, pigs, and llamas. Primary uses of hot water include washing animal-containment areas and laundry and showering facilities, each with a decentralized draining configuration. Solar hot water appears to be an appropriate renewable-energy technology to defray a portion of the hot water heating costs. This project will

• Research solar hot water heating technology

• Identify case studies of facilities that have installed this technology and research how the technology is actually working

- Determine cost and benefits at the Hospital for Large Animals
- Determine manufacturers and availability of equipment
- Identify installation and maintenance issues or benefits

Geothermal/Ground-Source Heat Pumps—Applicability on the Grafton Campus

Geothermal or ground-source heat pumps (GSHPs) are electrically powered systems that take advantage of the earth's relatively constant temperature to provide heating, cooling, and hot water for homes and commercial buildings. This project will

- Research GSHP technology
- Determine appropriate applications on the Grafton campus
- Identify case studies of facilities that have installed this technology and research how the technology is actually working
- Determine cost and benefits
- Determine manufacturers and availability of equipment
- Identify installation and maintenance issues or benefits

Heat Recovery at the Hospital for Large Animals

Health code regulations require an air exchange rate of fifteen total indoor air volumes per hour for the Hospital for Large Animals on the Grafton campus. Due to this requirement, there are several days of the year when the heating, ventilating, and air-conditioning (HVAC) system is required to cool the building during the day and heat the building at night. This project will

- Research HVAC heat-recovery technology
- Identify case studies of facilities that have installed this technology and research how the technology is actually working
- Determine costs and benefits at the Hospital for Large Animals
- Determine manufacturers and availability of equipment
- Identify installation and maintenance issues or benefits

Design of Solar Charging Station in the Buildings and Grounds Department—Medford Campus

The Buildings and Grounds Department of Tufts University purchased an electric mower to replace an aging four-cycle gasoline engine mower. In addition to significant reductions in noise pollution, this mower will greatly reduce air emissions, including CO_2, a major contributor to global climate change. However, air emissions are not eliminated completely. Air emissions are associated with the electricity used to recharge the batteries. A solar power charging station is proposed to further reduce emissions associated with the electric power necessary to charge the mower batteries. This project will

- Research solar electric charging technology
- Identify case studies of how the technology is actually working in similar applications
- Determine costs and benefits at the Buildings and Grounds Department
- Determine manufacturers and availability of equipment
- Identify installation and maintenance issues or benefits

Evaluate Fume-Hood Control Technology and Develop a Policy and Action Plan for Tufts

Fume hoods are a very complex issue. Numerous technical, logistical, and human health and safety issues need to be taken into consideration. A brochure on this topic would therefore require extensive research. The following is a partial list of all the topics to research:

- Research the energy consumption of fume hoods.
- Research the impact of fume hoods on heating and cooling.
- Research control mechanisms for fume hoods (e.g., Phoenix Controls).
- Research what other institutions have done on this issue (e.g., Brown and Harvard).
- Research the number of fume hoods in the United States/in New England. Research energy and money-saving examples that are non–Tufts specific.
- Use the section on climate change from the computer brochure.

Compile all the research into an extensive reference document that can be downloaded from the web.

Appendix F
Campus Trends: Green Campuses Get into Gear

Energy-Conserving Initiatives Take Off at Tufts, Elsewhere

Justin Feldman
Posted: 3/8/06

At Colleges Nationwide, Environmental Responsibility is a Priority

State-of-the-art facilities, highly talented student bodies, esteemed faculty and study abroad programs have long been components of colleges' marketability. Recently, however, another factor—the environmental sustainability or the "greenness" of the campus—was added to that list.

According to the United States Green Building Council, more then 110 colleges have either built structures that are environmentally friendly and meet high standards of energy efficiency, or are in the process of doing so.

Across the country, colleges and universities are showing an increasing responsibility toward the environment—starting with their own campuses. For example, at Middlebury College in Vermont, local forests supply wood for construction. At Carnegie Mellon in Pittsburgh, students study a new building's "living roof," which is meant to reduce storm water drainage and improve water quality.

At Tufts, Sophia Gordon Hall—the new, mostly-senior dormitory opening this coming fall—will be the University's first "green" building. That is, the dorm will be Tufts' first building constructed according to U.S. Green Building Council Leadership in Energy and Environmental Design (LEED) voluntary standards.

These parameters were developed in order to guide environmentally-friendly construction, and to encourage the use of energy-efficient technology and recycled and renewable construction materials.

Tufts Takes the LEED . . .

According to Associate Civil Engineering Professor Chris Swan, who taught a course last semester called "Engineering and the Construction Process" that studied the actual Tufts construction site, "you get certain points for doing environmentally conscious construction, and you get a certification when you [have enough points]. This will be Tufts' first LEED-accredited building."

There are many different aspects of LEED certification, depending on "where the materials come from and how you obtained them," Swan said. "All of that goes into the LEED certification process."

In 2002, the University received a $500,000 grant from the Massachusetts Renewable Energy Trust to ensure the installment of energy-saving features in all new buildings.

According to Sarah Hammond Creighton, the Tufts Climate Initiative (TCI) Project Manager, there will be several energy-efficient technologies included in Sophia Gordon Hall, but the "most visible to students and to members of the Tufts community will be two solar technologies."

One of these technologies is a photovoltaic rooftop, which will generate electricity for the building. The other is a solar thermal system that will use the sun's energy to heat water.

"Solar photovoltaic panels line some of the exterior of Sophia Gordon Hall—that's energy reduction—so that gets you points in the LEED accreditation process," Swan said.

These photovoltaic panels will resemble glass awnings and will offer shade, as well as convert energy from sunlight. "We expect the photovoltaic system will generate 26,000 kilowatt hours of electricity a year and we expect the solar thermal system will offset 30 percent to 40 percent, maybe even 50 percent of the energy used to heat water in the building," Creighton said.

The grant, however, does not cover most of the costs of these technologies, and the payoff will take years. "Payback is a good thing, but we also want to be good citizens . . . for the health of the planet," Creighton said.

Additionally, there is a storm water management plan in place on the site to control runoff and reduce the effects of erosion and sedimentation on the storm sewer system.

. . . and TCI Takes Action

While Sophia Gordon Hall is the most visible and publicized of the campus' green initiatives, TCI has in fact implemented multitudes of projects and programs that bring Tufts to the forefront of those institutions working to reduce greenhouse gas emissions and protect the environment.

"Tufts has been taking action to address energy issues for a long time for a couple reasons," Creighton said. "One is that it saves money and two is that Tufts has a commitment to civic engagement and to environmental stewardship."

This past year, TCI was one of only 17 recipients of a prestigious award from the Environmental Protection Agency (EPA) for its efforts to reduce climate change. According to the EPA Web site, other recipients of this honor include the cities of Syracuse, New York and Boulder, Colorado, the California Energy Commission and even McDonalds, Coca-Cola and Unilever Refrigerants Naturally Partnership.

American Electric Power, the largest energy generator and consumer of coal in the United States, also received the same honor for its efforts in preventing approximately 18 million cumulative tons of carbon dioxide from being produced and for its reforestation and conservation projects.

On campus, most of TCI's projects are in collaboration with groups like the Tufts Institute for the Environment, who provide much of the projects' funding, and Tufts Division of Operations, which eventually implements the projects.

Compact fluorescent bulbs, motion sensors, vending misers and front-loading washing machines are just a handful of the small-scale energy efficiency projects that TCI has been responsible for initiating.

"There are a lot of initiatives that have made a difference, from things that are viewable [such as] the solar panels on Fairmount House and the solar hot water on the French house to things that you don't see unless you look for them, like motion detectors in a lot of the buildings so the lights get turned off and save energy," explained Professor Ann Rappaport of the Urban and Environmental Policy and Planning School, who is also the TCI co-director.

Additionally, more energy-saving air conditioning systems and boilers with steam traps have been installed in buildings all over campus in order to reduce energy consumption. With these numerous energy saving projects, Creighton said proudly, "[Tufts is] on the cutting edge of this stuff."

"For all TCI projects, we work with people in facilities and in operations to look for ways to save energy and reduce emissions," Rappaport said. "One of the things that got done was [in] a renovation to the French house. There was a transition from heating oil to natural gas because natural gas produces a lot less climate-altering emissions than heating oil."

An Energy-Saving Switch

This winter, the efforts of the TCI and Tufts Energy Management lead Tufts to switch its electricity supplier. The new company, TransCanada, is located in Westborough, Massachusetts and supplies the University with hydroelectric power. The energy that comes from this new company is a mix of about 81 percent hydropower and 19 percent natural gas.

Previously, Tufts purchased its energy through a company that used only oil, natural gas and coal—sources that emit carbon dioxide and other green house gases that pollute the air. With this new source of clean hydroelectric power, greenhouse gas emissions connected to Tufts will be reduced to about 21 percent of the New England average.

"We look for multiple reasons for taking these initiatives. So, for example, something that will reduce emissions of climate-altering gas may also reduce costs to the University," Rappaport explained.

Thinking Globally and Locally

Additionally, the University has adopted international and local goals of keeping the air clean and reducing energy usage. In 1990, Tufts was the first University to sign the EPA Green Lights Pledge. This commitment was a promise that 90 percent of the lighting on campus would be upgraded more energy-efficient lighting.

"A lot of lighting that we had in 1990 you will not see at Tufts anymore," Creighton said.

The recent change in energy suppliers has aided Tufts in staying in line with the Kyoto Protocol, an objective aimed at reducing greenhouse gas emissions by seven percent of their 1990 levels between 2008 and 2012.

Tufts is also committed to the New England Governors' goal of reducing emissions by 10 percent by 2020, as well as the targets of the Chicago

Climate Exchange (which aims for one-percent reductions per year from 2003–2006).

Many proactive student groups around campus also do their part by creating awareness and outreach campaigns.

Environmental Consciousness Outreach (ECO) is one such student organization. In addition to the Do-It-In-The-Dark campaign in the fall, ECO is currently planning EarthFest, a week-long event in April to increase student awareness about environmental issues.

According to sophomore and ECO Officer Emma Shields, the ground is currently working on a "paper campaign."

"We're trying to get the departments to buy recycled paper," Shields said. "We are also working on a small campaign to get the fraternities to recycle. We're starting small and we've gotten the recycling bins to ZBT and AEPi yesterday. We're starting with the land that's owned by Tufts."

"We all live in this world, and we have to share it and keep it for the future generations," Shields added. "We can't be selfish. It's so easy to be environmental and take little steps."

"Tufts is doing really well compared to its peer institutions, and that is something we're really proud of," Creighton added.

Rebecca Dince contributed to this article.

Notes

Chapter 1

1. John Whitfield, "Alaska's Climate: Too Hot to Handle," *Nature* 425 (2003): 338–339.

2. National Oceanic and Atmospheric Administration, *Global Warming: Frequently Asked Questions* (online) (Asheville, NC: National Climatic Data Center, 2006), http://www.ncdc.noaa.gov/oa/climate/globalwarming.html.

3. National Oceanic and Atmospheric Administration, *Global Warming: Frequently Asked Questions.*

4. Mark Maslin, *Global Warming: A Very Short Introduction* (Oxford: Oxford University Press, 2004).

5. Intergovernmental Panel on Climate Change, *Climate Change 2001: Working Group II: Impacts, Adaptation and Vulnerability*, Section 1.1 "Overview of the Assessment" (New York: Cambridge University Press, 2001), www.grida.no/climate/ipcc_tar/wg2/005.htm.

6. See www.tufts.edu/climb and Paul Kirshen, *Tufts Civil Engineer Predicts Boston's Rising Sea Levels Could Cause Billions of Dollars in Damage* (online) (Medford, MA: Tufts Institute of the Environment, 2003), http://www.tufts.edu/tie/tci/pdf/Kirshen.pdf.

7. See for example P. Martens, "How Will Climate Change Affect Human Health?", *American Scientist* 87/6 (1999): 534. Also see United Nations Environment Programme, *Potential Impacts of Climate Change: Vital Climate Graphics* (online), 2006, http://www.grida.no/climate/vital/impacts.htm.

8. Ann Rappaport and John Blydenburgh, "Industrial Ecology in Large Companies: An Initial Assessment of Practice," unpublished manuscript.

9. Kay Lazar, "Expert: Business Shouldn't Ignore Global Warming," *Boston Herald,* January 14, 2003, 20.

10. CERES, *Value at Risk: Climate Change and the Future of Governance* (online) (Boston: CERES, 2002), http://www.ceres.org/reports/main.htm.

11. Andrea Lynn, *Americans Most Misinformed about Global Warming* (online) (Champaign, IL: *Inside Illinois*, 2002), http://www.news.uiuc.edu/II/03/0918/09globwarm_P.html.

Chapter 2

1. National Weather Service, *Climate Glossary* (online) (Camp Springs, MD: Climate Prediction Center, 2004), http://www.cpc.ncep.noaa.gov/products/outreach/glossary.shtml.

2. U. Siegenthaler et al., "Stable Carbon Cycle-Climate Relationship during the Late Pleistocene," *Science* 310 (2005): 1313–1317.

3. National Weather Service, *Climate Glossary* (online) (Camp Springs, MD: Climate Prediction Center, 2004), http://www.cpc.ncep.noaa.gov/products/outreach/glossary.shtml.

4. R. Watson et al., *Climate Change 2001: The Scientific Basis*, Intergovernmental Panel on Climate Change (Cambridge: Cambridge University Press, 2001); emphasis added.

5. R. Watson et al., *Climate Change 2001*.

6. Stephen Schneider et al., eds., *Climate Change Policy: A Survey* (Washington, DC: Island Press, 2002), 15–16.

7. John D. Sterman and Linda Booth Sweeney, "Cloudy Skies: Assessing Public Understanding of Global Warming," *System Dynamics Review* 18/2 (2002): 207–240.

8. Sterman and Sweeney, "Cloudy Skies."

9. Sterman and Sweeney, "Cloudy Skies," 208.

10. See IPCC's website, www.ipcc.ch.

11. Union of Concerned Scientists, *Key Findings of Working Group II: "Climate Change 2001: Impacts, Adaptation and Vulnerability"* (online), 2001, http://www.ucsusa.org/global_environment/global_warming/page.cfm?pageID=521.

12. Union of Concerned Scientists, *Key Findings of Working Group II*.

13. Union of Concerned Scientists, *Key Findings of Working Group II*.

14. "Negotiations on what was to become the Kyoto Protocol were launched by the Conference of the Parties (COP) at its first session (Berlin, March/April 1995) when it adopted its decision 1/CP.1 (the "Berlin Mandate")" Joanna Depledge, *Tracing the Origins of the Kyoto Protocol: An Article-by-Article Textual History* (online) (Bonn: UNFCCC, 2000), http://unfccc.int/resource/docs/tp/tp0200.pdf.

15. Anil Agarwal, "A Southern Perspective on Curbing Global Climate Change," in Stephen Schneider et al., eds., *Climate Change Policy: A Survey* (Washington, DC: Island Press, 2002).

16. G. Marland, T. A. Boden, and R. J. Andres, *Global, Regional, and National CO_2 Emissions. In Trends: A Compendium of Data on Global Change* (Oak Ridge, TN: Carbon Dioxide Information Analysis Center, Oak Ridge National Laboratory, U.S. Department of Energy, 2003), http://cdiac.esd.ornl.gov/trends/emis/em_cont.htm.

17. Environmental Protection Agency, *Climate Leaders* (online) (Washington, DC: Environmental Protection Agency, 2004), http://www.epa.gov/climateleaders/.

18. International Council for Local Environmental Initiatives, "US Cities for Climate Protection Campaign (CCP)" (online) (Berkeley, CA: International Council for Local Environmental Initiatives, 2004), http://www.iclei.org/us/ccp/.

19. Andrew Revkin, "New York City and 8 States Plan to Sue Power Plants," *New York Times*, July 21, 2004, section A, p. 15.

20. The Committee on the Environment and the Northeast International Committee on Energy of the Conference of New England Governors and Eastern Canadian Premiers, *Climate Change Action Plan 2001* (Boston: New England Governors/Eastern Canadian Premiers, August 28, 2001), http://www.negc.org/documents/NEG-ECP%20CCAP.PDF.

Chapter 3

1. Thomas Gloria, *Tufts University's Greenhouse Gas Emissions Inventory for 1990 and 1998* (Medford, MA: Tufts University, 2001).

2. See, for example, Clean Air/Cool Planet (www.cleanair-coolplanet.org); Torrie Smith Associates, eMission software (www.torriesmith.com); Environmental Software Providers (ESP) Greenhouse Gas Suite TM (www.esp-net.com).

3. Rocky Mountain Institute, *Tunneling through the Cost Barrier: Why Big Savings Often Cost Less Than Small Ones* (online) (Aspen, CO: Rocky Mountain Institute, summer 1997), http://www.rmi.org/images/other/Newsletter/NLRMIsum97.pdf.

4. Erica Noonan, "Wellesley Gift to Generate Years of Buzz: Alumna Leaves $27 m, Most for Power Plant, "*Boston Globe*, May 20, 2005, A1.

5. Massachusetts Sustainability Office, *UMass Boston Energy Retrofit Project* (online) (Boston: Department of Environmental Protection, 2006), http://www.mass.gov/envir/Sustainable/initiatives/initiatives_GHG.htm#umbretrofits.

6. A. B. Rappaport and S. H. Creighton, "Effective Campus Environmental Assessment," *Planning for Higher Education* 31/3 (2003): 45–52.

Chapter 4

1. Thomas Gold (online), *The Economist* (London); July 1, 2004, http://www.economist.com/people/PrinterFriendly.cfm?Story_ID=2876655.

2. See, for example, the experiences of Connecticut College and the University of Colorado at Boulder.

3. See http://www.rso.cornell.edu/kyotonow/index.html.

4. See http://www.startingbloc.org/.

5. See http://campusclimatechallenge.org/.

Chapter 5

1. Lights of America, package information on model 2509 compact fluorescent bulb, 2001.

2. See, for example, Frank Ackerman, *Why Do We Recycle?* (Washington, DC: Island Press, 1997).

3. Society of Environmental Toxicology and Chemistry, *Life-Cycle Assessment (LCA)* (online) (Pensacola, FL: Professional Interest Groups, 2006), http://www.setac.org/htdocs/who_intgrp_lca.html.

4. CERES, *Value at Risk: Climate Change and the Future of Governance* (online) (Boston: CERES, 2002), http://www.ceres.org/reports/main.htm.

5. National Association of Energy Service Companies, *What Is an ESCO?* (online) (Washington, DC: NAESCO, 2006), http://www.naesco.org/about/esco.htm.

Chapter 6

1. Pew Center for Global Change, *Global Warming Basics: Executive Summary* (online) (Arlington, VA: Pew Center for Global Change, undated), http://www.pewclimate.org/global-warming-in-depth/all_reports/buildings/ex__summary.cfm.

2. U.S. Environmental Protection Agency. *EPA's Draft Report on the Environment 2003* (online) (Washington, DC: U.S. Environmental Protection Agency, http://www.epa.gov/indicators/roe/pdf/tdAir1-3.pdf.

3. New Buildings Institute, *Integrated Energy Systems: Productivity and Building Science: Final Report* (online) (White Salmon, WA: New Buildings Institute, 2003), http://www.newbuildings.org/downloads/FinalAttachments/PIER_Final_Report(P500-03-082).pdf.

4. Joseph Romm and William Browning, *Greening the Building and the Bottom Line: Increasing Productivity through Energy-Efficient Design* (Aspen, CO: Rocky Mountain Institute, 1994).

5. *Bowdoin College, Bowdoin Works Hard to Make Sustainable Living Decisions a Reality* (online) (Brunswick, ME: Bowdoin College, 2003), http://www.bowdoin.edu/news/archives/1bowdoincampus/001330.shtml.

6. Keisha Payson, Sustainability Coordinator at Bowdoin College, personal communication.

7. Stephanie Ebbert, "Wind Turbines Gaining Power: Smaller Communities, Colleges Plan Projects," *Boston Globe,* February 24, 2006, http://www.mass.gov/envir/Sustainable/pdf/vf05 mma windturbine.pdf.

8. MA DCAM, *Division of Capital Asset Management Energy Efficiency and Sustainable Design Program Highlights: A Contribution to the Mayor's Green Building Task Force* (online) (Boston: MA DCAM, December 18, 2003), http://www.cityofboston.gov/bra/gbtf/documents/DCAM-program03-12-18.PDF.

9. William R. Moomaw, "Aligning Values for Effective Sustainability Planning," *Planning for Higher Education* 31/3 (2003): 159–164 (quote on 160).

10. William Browning, "Successful Strategies for Planning a Green Building," *Planning for Higher Education* 31/3 (2003): 114.

11. See http://www.energystar.gov/index.cfm?c=new_bldg_design.bus_target_finder.

12. Jason F. McLennan and Peter Rumsey, "*Green Edge" in Environmental Design and Construction* (online), September 2003, http://www.edcmag.com/CDA/ArticleInformation/features/BNP__Features__Item/0,4120,107521,00.html.

13. Interview with author, August 2005.

14. McLennan and Rumsey, "*Green Edge*."

15. Amory Lovins, "Canada's Energy Efficiency Conference, "Ottawa, May 19, 1999, http://oee.nrcan.gc.ca/headsup/archives/jun_1999.cfm?PrintView=N&Text=N.

16. William Leahy, Associate Executive Director of Operation at the Institue for Sustainable Energy at Eastern Connecticut State University, March 13, 2006, personal communication.

17. State University of New York Buffalo. *UB Green* (online), undated, http://www.wings.buffalo.edu/ubgreen/.

Chapter 7

1. Vermont Law School, *Learning by Doing: James L. and Evelena S. Oakes Hall Environmental Features* (online) (Royalston, VT: Vermont Law School, undated), http://www.vermontlaw.edu/life/index.cfm?doc_id=559.

2. Donald Wulfinghoff, *Energy Efficiency Manual* (Wheaton, MD: Energy Institute Press, 1999), 1017.

3. Lawrence Berkeley National Laboratory, Arkwright Mutual Insurance Co., U.S. DOE, and Arkwright Education, *Campus Lighting: Lighting Efficiency Options for Student Residential Living Units* (online), 2004, http://eetd.lbl.gov/EMills/PUBS/PDF/Hal_Fact2.pdf.

4. DOE Energy Efficiency and Renewable Energy, *Industrial Technologies Program Best Practices: Collaboration Advances Motor Efficiency* (online) (Washington, DC: DOE Energy Efficiency and Renewable Energy, 2006), http://www1.eere.energy.gov/industry/bestpractices/wint2002_collaboration.html.

5. Wulfinghoff, *Energy Efficiency Manual.*

6. Alan Bandes and Bruce Gorelick, *Inspect Steam Traps for Efficient System* (online) (Terre Haute, IN: TWI Press, Inc., 2000), http://www.maintenanceresources.com/ReferenceLibrary/SteamTraps/Inspect.htm.

7. Energy Design Resources, *Design Briefs: Energy Management Systems* (online) (CA: Energy Design Resources, undated), http://www.energydesignresources.com/resource/18/.

8. See http://www.labs21century.gov.

9. U.S. EPA and NREL, *Laboratories for the 21st Century: An Introduction to Low Energy Design* (online) (Washington, DC: U.S. EPA and NREL, 2000), http://www.labs21century.gov/pdf/lowenergy_508.pdf.

10. Scott Taylor, *Fume Hood Study: Tufts University*, 2004, unpublished.

11. Taylor, *Fume Hood Study.*

12. Neil Rasmussen, American Power Conversion, *Avoidable Mistakes That Compromise Cooling Performance in Data Centers and Network Rooms, White Paper #49* (online) (W. Kingston, RI: American Power Conversion, 2003), http://www.apcmedia.com/salestools/SADE-5TNRKG_R0_EN.pdf.

13. Tony Evans, American Power Conversion, *The Different Types of Air Conditioning Equipment for IT Environments, White Paper #59* (online) (W. Kingston, RI: American Power Conversion, 2004), http://www.apcmedia.com/salestools/VAVR-5UDTU5_R0_EN.pdf.

14. Wulfinghoff, *Energy Efficiency Manual.*

15. Wulfinghoff, *Energy Efficiency Manual.*

16. Oak Ridge National Laboratory, *Guide to Combined Heat and Power Systems for Boiler Owners and Operators* (ORNL/TM-2004/144) (Oak Ridge, TN: Oak Ridge National Laboratory, 2004), http://www1.eere.energy.gov/industry/bestpractices/pdfs/guide_chp_boiler.pdf.

17. Federal Energy Management Program, *FEMP CHP Program Overview* (online) (Washington, DC: Federal Energy Management Program, undated), http://www.eere.energy.gov/femp/pdfs/chp_prog_overvw.pdf.

18. For a listing of university CHP applications throughout the United Stated and Canada, see http://www.energy.rochester.edu/us/list.htm.

19. See http://www.eere.energy.gov/states/.

20. Emory University, *Emory Research Building First in Southeast to Be Certified by U.S. Green Building Council* (online) (Atlanta: Emory University, September 6, 2002), http://news.emory.edu/Releases/LEED1031342654.html.

21. University of California Office of Strategic Communications, *Green Building Policy and Clean Energy Standards* (online), August 2003, http://www.ucop.edu/news/factsheets/greenbuildings.pdf.

22. LEED EB website, http://www.usgbc.org/DisplayPage.aspx?CMSPageID=221&.

23. Bill Von Neida and Tom Hicks, U.S. Environmental Protection Agency, *Building Performance Defined: The ENERGY STAR National Energy Performance Rating System* (online) (Washington, DC: U.S. Environmental Protection Agency, undated), http://www.energystar.gov/ia/business/tools_resources/aesp.pdf.

24. See https://www.energystar.gov/istar/pmpam/.

25. Von Neida and Hicks, *Building Performance Defined.*

Chapter 8

1. Kevin Lyons, *Buying for the Future: Contract Management and the Environmental Challenge* (Ann Arbor: University of Michigan Press, 1999).

2. See http://www.mass.gov/?pageID=osdmodulechunk&L=1&L0=Home&sid=Aosd&b=terminalcontent&f=osd_es_green&csid=Aosd.

3. G. Norris, *Direct and Upstream Emissions of Carbon Dioxide from Fossil Fuel Combustion*, report submitted to the National Renewable Energy Laboratory (North Berwick, ME: Sylvatica, February 1998).

4. Thomas Gloria, unpublished letter to Environmental Defense, March 2001.

5. U.S. Environmental Protection Agency, *What Is Green Power?* (online) (Washington, DC: U.S. Environmental Protection Agency, 2006), http://www.epa.gov/greenpower/whatis/intro.htm.

6. U.S. Department of Energy, *Mercury Emission Control R&D* (online) (Washington, DC: U.S. Department of Energy, 2006), www.fossil.energy.gov/programs/powersystems/pollutioncontrols/overview_mercurycontrols.html.

7. Environmental News Service, *Global Wind Power Industry Spins into High Gear* (online), 2006, www.ens-newswire.com/ens/feb2006/2006-02-23-04.asp.

8. U.S. Environmental Protection Agency, *Guide to Purchasing Green Power* (online) (Washington, DC: U.S. Environmental Protection Agency, 2004), www.epa.gov/greenpower/pdf/purchasing_guide_for_web.pdf.

9. U.S. Environmental Protection Agency, *Guide to Purchasing Green Power*.

10. U.S. Environmental Protection Agency, *Green Power 2004 Award Winners* (online) (Washington, DC: U.S. Environmental Protection Agency, 2004), http://www.epa.gov/greenpower/winners/2004_awards.htm#college.

11. Western Washington University, *WWU Receives National Green Power Award: Western Honored for Environmental Commitment, Student Initiative* (online) (Bellingham: Western Washington University, 2005), http://west.wwu.edu/ucomm_news/articles/988.asp.

12. U.S. Environmental Protection Agency, *Green Power Partners*. (online) (Washington, DC: U.S. Environmental Protection Agency, 2005), www.epa.gov/grnpower/partners/partners/universityofpennsylvania.htm.

13. *Tufts Climate Initiative* (online) (Medford, MA: Tufts University, undated), http://www.tufts.edu/tie/tci/WashingMachines.html.

14. See http://www.tufts.edu/tie/tci/VendingMisers.html.

15. Energy Star Program, "Office Equipment" (online) (Washington, DC: Energy Star Program, undated), http://www.energystar.gov/index.cfm?c=ofc_equip.pr_office_equipment.

16. Energy Star Program, "Buy Products That Make a Difference" (online) (Washington, DC: U.S. EPA Energy Star Program, undated), http://www.energystar.gov/index.cfm?fuseaction=find_a_product.

17. William Cowart and Veronika Pesinova, *An Assessment of GHG Emissions from the Transportation Sector* (online) (Washington, DC: U.S. EPA 2003), http://www.epa.gov/ttn/chief/conference/ei12/green/pesinova.pdf.

18. Donald Shoup, in Will Toor and Spenser Havlick, *Transportation and Sustainable Campus Communities* (Washington, DC: Island Press, 2004), 81.

19. Melissa Burden, "Big Ten Student Bus Fees, Services Vary by University" (online) (Ann Arbor: University of Michigan: 1999), http://www.statenews.com/editionsspring99/022399/p1_cata.html.

20. North Carolina State University. "Transportation" (online) (Raleigh, NC: North Carolina State University, 2005), http://www2.acs.ncsu.edu/trans/transit/upass.html.

21. National Biodiesel Board, *Environmental Benefits.* (online) (Jefferson City, MO: National Biodiesel Board, 2004), http://www.biodiesel.org/pdf_files/Enviro_Benefits.PDF.

22. Beth, Potier, "Goodbye Black Smoke, Hello Green Transit," *Harvard University Gazette*, March 4, 2004. http://www.hno.haward.edu/gazette/2004/03.04/07-biodiesel.html.

23. Kathleen Burge, "Ice and Isolation for Nantucket: Severe Freeze Pinches Supplies and Island Residents' Sense of Freedom," *Boston Globe*, January 27, 2004, section Metro, A1.

24. National Biodiesel Board. *Environmental Benefits.*

25. Harry, Rijnen, "Offsetting Environmental Damage by Planes," *New York Times*, February 18, 2003, C6.

26. Thomas Gloria, unpublished manuscript.

27. Rijnen, "Offsetting Environmental Damage by Planes."

28. U.S. Energy Information Administration, "Technical Assistance: General Guidance, Project-Specific Guidance, Reporting Tools, and Emission Factors" (online) (Washington, DC: U.S. Energy Information Administration, undated), http://www.eia.doe.gov/oiaf/1605/techassist.html.

29. See http://www.dumpandrun.org.

30. Energy Star Program, "Restaurants" (online) (Washington, DC: Energy Star Program, undated), http://www.energystar.gov/index.cfm?c=small_business.sb_restaurants.

31. U.S. Environmental Protection Agency, "Methane: Sources and Emissions" (online) (Washington, DC: U.S. Environmental Protection Agency, 2006), http://www.epa.gov/methane/sources.html.

32. U.S. Environmental Protection Agency, "Technical Approaches and Policy Options for Reducing Greenhouse Gas Emissions" (online) (Washington, DC: U.S. Environmental Protection Agency, undated), http://yosemite.epa.gov/OAR/globalwarming.nsf/UniqueKeyLookup/SHSU5BVKEX/$File/part-2.pdf.

33. Ibid.

Chapter 9

1. See, for instance, www.iclei.org/index.php?id=1118 and http://yosemite.epa.gov/oar/globalwarming.nsf/content/ActionsStateActionPlans.html.

2. F. Jacobs and J. Kapuscik, *Making It Count: Evaluating Family Preservation Services* (Medford, MA: Tufts University, 2000).

3. The program has expanded beyond Burlington. See http://www.10percentchallenge.org/.

4. UNEP Finance Initiatives Climate Change Working Group, *CEO Briefing* (online) (New York: UNEP Finance Initiatives Climate Change Working Group, undated), http://www.innovestgroup.com/pdfs/2003-11_ceobriefing.pdf.

5. UNEP, *Global Environmental Outlook, Chapter 2: Natural Disasters* (online) (New York: UNEP, 2000), http://www.unep.org/geo2000/English/0038.htm#img31a.

6. Kay Lazar, "Expert: Business Shouldn't Ignore Global Warming," *Boston Herald*, January 14, 2003, 20.

7. See www.tufts.edu/climb and "Tufts Civil Engineer Predicts Boston's Rising Sea Levels Could Cause Billions of Dollars in Damage," March 2003, http://www.tufts.edu/tie/tci/pdf/Kirshen.pdf.

8. Brown University News Service, *Corporation Adopts Strategic Framework for Physical Planning* (online) (Providence, RI: Brown University, October 11, 2003), http://www.brown.edu/Administration/News_Bureau/2003-04/03-039.html.

9. Middlebury College, *Environmental Studies and Awareness: An Environmental Peak of Excellence* (online) (Middlebury, VT: Middlebury College, undated), http://www.middlebury.edu/offices/enviro/initiatives/environmental_peak.htm.

10. Johns Hopkins University, *A Plan for the Homewood Campus, Principles* (online) (Baltimore: Johns Hopkins University, 2000), http://www.jhu.edu/masterplan/principles.

11. Brandeis University, *Campus Master Planning Project* (online) (Waltham, MA: Brandeis University, undated), http://www.brandeis.edu/masterplan/.

12. University of Chicago, *Campus Utilities* (online) (Chicago: University of Chicago, undated), http://www.uchicago.edu/docs/mp-site/masterplan/c7infrastruc/c7z-utilities-a.html.

13. Graham Jones, *World Oil and Gas "Running Out,"* (online), October 2, 2003, http://edition.cnn.com/2003/WORLD/europe/10/02/global.warming/.

14. Larry Goldstein, *Behind the Scenes in the University Boardroom: How Decision Makers Assess Facility Expenditures* (online), Agile Planner Workshop, October 16, 2003, http://www.agileplanner.org/pdf/Goldstein_Fall2003.pdf.

15. CERES, "Value at Risk: Climate Change and the Future of Governance" (online) (Boston: CERES, April 2002), http://www.ceres.org/pdf/climate.pdf.

16. UNEP Finance Initiatives Climate Change Working Group, undated.

17. CERES, "Value at Risk."

18. Kevin Morrison, "European Index Marks New Era in Investing," *Financial Times* (London), June 22, 2004, 47.

19. Joseph J. Romm, *Cool Companies* (Washington, DC: Island Press, 1999).

20. http://ceres.org/our_work/sgp.htm.

21. "Memorandum to Faculty in Arts, Sciences and Engineering on Budget Challenges," February 23, 2006.

22. Romm, *Cool Companies*, 47.

23. Eric A. Kriss, *Use of State Vehicles by Executive Agencies*, Commonwealth of Massachusetts ANF Bulletin 10, December 1, 2003.

Chapter 10

1. Press release, University of Illinois at Urbana-Champaign, "Americans among Most Misinformed about Global Warming," http://www.eurekalert.org/pub_releases/2003-09/uoia-aam090303.php. For the journal article on the study, see "Comparative Public Opinion and Knowledge on Global Climatic Change and the Kyoto Protocol: The U.S. versus the World?", *International Journal of Sociology and Social Policy* 23/1 (2003): 106–134.

2. John D. Sterman and Linda Booth Sweeney, "Cloudy Skies: Assessing Public Understanding of Global Warming," *System Dynamics Review* 18/2 (quote on 233) (2002): 207–240.

3. John D. Sterman and Linda Booth Sweeney, "Understanding Public Complacency about Climate Change: Adults' Mental Models of Climate Change Violate Conservation of Matter," working paper, undated, http://web.mit.edu/jsterman/www/.

4. W. Kempton, J. S. Boster, and J. A. Hartley, *Environmental Values in American Culture* (Cambridge, MA: MIT Press, 1995).

5. Rawle O. King, "Hurricane Katrina: Insurance Losses and National Capacities for Financing Disaster and Risk," Congressional Research Service, Library of Congress, Order Code RL33086, September 15, 2005, http://fpc.state.gove/documents/organization/53686.pdf.

6. UNEP/GRID-Arendal-Climate Change, http://www.grida.no/activities.cfm?pageID=2.

7. Hadley Centre, Climate Information, http://www.meto.gov.uk/research/hadleycentre/models/climate_system.html.

8. U.S. Environmental Protection Agency, Global Warming website, http://yosemite.epa.gov/oar/globalwarming.nsf/content/index.html.

9. Energy Information Administration, Household Vehicle Energy Consumption Report Series: DOE/EIA-0464, http://www.eia.doe.gov/emeu/rtecs/chapter3.html.

10. http://yosemite.epa.gov/oar/globalwarming.nsf/content/ResourceCenterToolsGHGCalculator.html.

11. See, for example, http://www.pollingreport.com/enviro.htm.

12. James Hannah, "As Students Use More Power, Colleges Must Rewire Rooms," Associated Press, November 20, 2003, http://msnbc.msn.com/id/3607032/.

13. Hannah, "As Students Use More Power, Colleges Must Rewire Rooms."

14. http://www.rowan.edu/studentaffairs/reslife/welcomeback/move_in_day/.

15. Greg Winter, "Jacuzzi U? A Battle of Perks to Lure Students," *New York Times*, October 5, 2003, 1.

16. Walter C. Swap, "Psychological Factors in Environmental Decision Making: Social Dilemmas," in R. A. Chechile and S. Carlisle, eds., *Environmental Decision Making: A Multidisciplinary Perspective* (New York: Van Nostrand Reinhold, 1991).

17. See, for example, W. Kempton, J. S. Boster, and J. A. Hartley, *Environmental Values in American Culture* (Cambridge, MA: MIT Press, 1995).

18. For a concise review of social marketing literature, see Kristin Marcell, Julian Agyeman, and Ann Rappaport, "Cooling the Campus: Experiences from a Pilot Study to Reduce Electricity Use at Tufts University, USA, Using Social Marketing Methods," *Sustainability in Higher Education* 5/2 (2004): 169–189.

19. Marcell, Agyeman, and Rappaport, "Cooling the Campus."

20. D. Meadows, "Places to Intervene in a System (in Increasing Order of Effectiveness)," *Whole Earth*, winter 1997, 78–84.

21. Kurt Teichert, panelist, Agile Planner Workshop, Boston, October 26, 2004.

22. Ross Gelbspan, *The Heat Is On* (Reading, MA: Perseus, 1998).

23. Ross Gelbspan, *Boiling Point* (New York: Basic Books, 2004).

24. According to http://www.uoregon.edu/~ecostudy/slp/energy/Energyuse.html, University of Oregon students are paying a $20/term electricity surcharge, which covers more than half the electric bill. Most schools in the Oregon system had an energy surcharge (http://www.ous.edu/news/press/072001a.htm).

25. A state judge ruled that the University of Washington did not have the authority to charge an energy fee because the fee constituted a tuition increase and only the legislature is authorized to increase tuition at state institutions. See Ben Gose, "Students in Washington Win Energy-Fee Dispute," *Chronicle of Higher Education*, November 30, 2001, 21.

Chapter 11

1. The Piedmont Project, http://www.scienceandsociety.emory.edu/piedmont/.

2. Kristin Marcell, Julian Agyeman, and Ann Rappaport, "Cooling the Campus: Experiences from a Pilot Study to Reduce Electricity Use at Tufts University, USA, Using Social Marketing Methods," *Sustainability in Higher Education* 5/2 (2004): 187.

3. http://www.usgcrp.gov/usgcrp/Library/ocp2004-5/default.htm.

4. For the report from the project *Climate's Long-Term Impacts on Metro Boston*, see http://www.tufts.edu/tie/climb/.

Chapter 12

1. Kresge Foundation, http://www.kresge.org/.

2. Massachusetts Technology Collaborative, http://www.mtpc.org/.

3. Energy Star, http://www.energystar.gov/index.cfm?fuseaction=find_a_product.

Bibliography

Ackerman, Frank. 1997. *Why Do We Recycle?* Washington, DC: Island Press.

Alley, Richard. 2002. *The Two-Mile Time Machine: Ice Cores, Abrupt Climate Change, and Our Future*. Princeton, NJ: Princeton University Press.

Anink, David. 1996. *Handbook of Sustainable Building: An Environmental Preference Method for Selection of Materials for Use in Construction and Refurbishment*. London: James & James.

Barlett, Peggy, and Gregory Chase, eds. 2004. *Sustainability on Campus: Stories and Strategies for Change*. Cambridge, MA: MIT Press.

Bearg, David W. 1993. *Indoor Air Quality and HVAC Systems*. Boca Raton, FL: CRC Press.

Begg, Kathryn, Frans van der Woerd, and David Levy. 2005. *The Business of Climate Change: Corporate Responses to Kyoto*. Sheffield, UK: Greenleaf.

Blatt, Harvey. 2004. *America's Environmental Report Card: Are We Making the Grade?* Cambridge, MA: MIT Press.

Brower, Michael, and Warren Leon. 1999. *The Consumer's Guide to Effective Environmental Choices*. New York: Three Rivers Press.

Burton, Ian, Elizabeth Malone, and Saleemul Huq. 2005. *Adaptation Frameworks for Climate Change: Developing Strategies, Policies and Measures*. New York: United Nations Development Programme/Cambridge University Press.

Carmody, John, Stephen Selkowitz, and Lisa Heschong 1996. *Residential Windows*. New York: Norton.

Chechile, R. A., and S. Carlisle, eds. 1991. *Environmental Decision Making: A Multidisciplinary Perspective*. New York: Van Nostrand Reinhold.

Claussen, Eileen, Vicki Arroyo Cochran, and Debra P. Davis, eds. 2002. *Climate Change: Science, Strategies, and Solutions*. Boston: Brill

Creighton, Sarah Hammond. 1998. *Greening the Ivory Tower: Improving the Environmental Track Record of Universities, Colleges, and Other Institutions*. Cambridge, MA: MIT Press.

Crosbie, Michael J., Steven Winter Associates, Inc., ed. 1997. *The Passive Solar Design and Construction Handbook*. Hoboken, NJ: Wiley.

Dauncey, Guy, and Patrick Mazza. 2001. *Stormy Weather: 101 Solutions to Global Climate Change*. Gabriola Island, BC, Canada: New Society Publishers.

DeCanio, Stephen. 2003. *Economic Models of Climate Change: A Critique*. New York: Palgrave Macmillan.

Dernback, John, ed. 2002. *Stumbling toward Sustainability.* Washington, DC: Environmental Law Institute.

Flannery, Tim. 2005. *The Weather Makers: How Man Is Changing the Climate and What It Means for Life on Earth*. New York: Atlantic Monthly Press.

Freestone, David, and Charlotte Streck, eds. 2005. *Legal Aspects of Implementing the Kyoto Protocol Mechanisms: Making Kyoto Work*. New York: Oxford University Press.

Gelbspan, Ross. 1998. *The Heat Is On: The Climate Crisis, the Cover-Up, the Prescription*. Reading, MA: Perseus Books.

Gelbspan, Ross. 2004. *Boiling Point*. New York: Basic Books.

Hall, C. Michael, and James Higham, eds. 2005. *Tourism, Recreation and Climate Change*. Buffalo: Channel View.

Helm, Dieter, ed. 2005. *Climate-Change Policy*. Oxford: Oxford University Press.

Houghton, John T. 2004. *Global Warming: The Complete Briefing*. New York: Cambridge University.

Houghton, J., Y. Ding, D. J. Griggs, M. Noguer, P. J. van der Linden, X. Dai, K. Maskell, and C. A. Johnson, eds. 2001. *Climate Change 2001: The Scientific Basis*. Contribution of Working Group I to the Third Assessment Report of the Intergovernmental Panel on Climate Change. Cambridge: Cambridge University Press.

Jacobs, Francine, and Jennifer Kapuscik. 2000. *Making It Count: Evaluating Family Preservation Services*. Medford, MA: Tufts University.

Kempton, W., J. S. Boster, and J. A. Hartley. 1995. *Environmental Values in American Culture*. Cambridge, MA: MIT Press.

Keniry, Julian. 1995. *Ecodemia: Campus Environmental Stewardship at the Turn of the 21st Century.* Washington, DC: National Wildlife Federation.

Kolbert, Elizabeth. 2006. *Field Notes from a Catastrophe*. London: Bloomsbury.

Komp, Richard. 2002. *Practical Photovoltaics: Electricity from Solar Cells*. Ann Arbor, MI: Aatec Publications.

Linden, Eugene. 2006. *The Winds of Change: Climate, Weather, and the Destruction of Civilizations*. New York: Simon & Schuster.

Lovins, Amory B. 2004. *Winning the Oil Endgame*. Snowmass, CO: Rocky Mountain Institute.

Lynas, Mark. 2004. *High Tide: the Truth about Our Climate Crisis*. New York: Picador.

Lyons, Kevin. 1999. *Buying for the Future: Contract Management and the Environmental Challenge*. Ann Arbor: University of Michigan Press.

Markvart, Tomas. 2000. *Solar Electricity*. 2nd ed. Hoboken, NJ: Wiley.

Masika, Rachel. 2002. *Gender, Development and Climate Change*. Oxford: Oxfam Publishing.

Maslin, Mark. 2004. *Global Warming: A Very Short Introduction*. Oxford: Oxford University Press.

McDonough, William, and Michael Braungart. 2003. *The Hannover Principles: Design for Sustainability.* Charlottesville, VA: William McDonough and Partners.

Moll, Gary, and Sara Ebenreck, eds. 1989. *Shading Our Cities: A Resource Guide for Uban and Community Forest*. Washington, DC: Island Press.

O'Hare, Greg, John Sweeney, and Rob Wilby. 2005. *Weather, Climate, and Climate Change: Human Perspectives*. New York: Pearson Prentice Hall.

Orr, David W. 1992. *Ecological Literacy: Education and the Transition to a Postmodern World.* Albany: State University of New York Press.

Orr, David W. 2002. *The Nature of Design: Ecology, Culture and Human Intention.* Oxford: Oxford University Press.

Orr, David W. 2004. *Earth in Mind: On Education, Environment, and the Human Prospect.* Washington, DC: Island Press.

Rabe, Barry. 2004. *Statehouse and Greenhouse: The Emerging Politics of American Climate Change Policy*. Washington, DC: Brookings Institution Press.

Ravindranath, N. H., and Jayant A. Sathaye. 2002. *Climate Change and Developing Countries*. Boston: Kluwer.

Roaf, Sue, David Crichton, and Fergus Nicol. 2005. *Adapting Buildings and Cities for Climate Change*. Oxford: Architectural Press.

Romm, Joseph J. 1999. *Cool Companies*. Washington, DC: Island Press.

Romm, Joseph J. 2004. *The Hype about Hydrogen: Fact and Fiction in the Race to Save the Climate.* Washington, DC: Island Press.

Schneider, Stephen H., and Terry L. Root. 2002. *Wildlife Responses to Climate Change: North American Case Studies*. Washington, DC: Island Press.

Schneider, Stephen H., Armin Rosencranz, and John O. Niles, eds. 2002. *Climate Change Policy: A Survey*. Washington, DC: Island Press.

Speth, James G. 2004. *Red Sky at Morning: America and the Crisis of the Global Environment*. Chicago: R. R. Donnelly & Sons.

Spiegel, Ross, and Dru Meadows. 2006. *Green Building Materials: A Guide to Product Selection and Specification*. 2nd ed. Hoboken, NJ: Wiley.

Suozzo, Margaret, Jim Benya, Mark Hydeman, Paul Dupont, Steven Nadel, and R. Neal Elliott. 2002. *Guide to Energy-Efficient Commercial Equipment*. 2nd ed. Washington, DC: American Council for an Energy Efficient Economy.

Sustainable Buildings Industry Council. 2002. *Green Buildings Guidelines*. Washington, DC: SBIC.

Swisher, Joel. 2003. *The New Business Climate: A Guide to Lower Carbon Emissions and Better Business Performance*. Snowmass, CO: Rocky Mountain Institute.

Toor, Will, and Spenser Havlick. 2004. *Transportation and Sustainable Campus Communities*: Issues, Examples, Solutions. Washington, DC: Island Press.

Townsend, A. K. 1997. *The Smart Office: Turning Your Company on Its Head.* Olney, MD: Gila Press.

Wilson, Alex, Jennifer Thorne, and John Morrill. 2004. *Consumer Guide to Home Energy Savings*. 8th ed. Washington, DC: American Council for an Energy Efficient Economy.

Wilson, Alex, and Nadav Malin, eds. 2005. *GreenSpecDirectory*. 5th ed. Montpelier, VT: Capital City Press.

Wulfinghoff, Donald. 1999. *Energy Efficiency Manual*. Wheaton, MD: Energy Institute Press.

Index